Analog Circuits and Signal Processing

Series Editors

Mohammed Ismail, The Ohio State University
Mohamad Sawan, École Polytechnique de Montréal

For further volumes:
http://www.springer.com/series/7381

Jens Masuch · Manuel Delgado-Restituto

Ultra Low Power Transceiver for Wireless Body Area Networks

 Springer

Jens Masuch
Manuel Delgado-Restituto
Institute of Microelectronics of Sevilla
University of Sevilla
Sevilla
Spain

ISBN 978-3-319-03367-9   ISBN 978-3-319-00098-5 (eBook)
DOI 10.1007/978-3-319-00098-5
Springer Cham Heidelberg New York Dordrecht London

Printed on acid-free paper

Springer is part of Springer Science+Business Media (www.springer.com)

Preface

Wireless Body Area Networks (WBANs) are expected to promote new applications for the ambulatory health monitoring of chronic patients and the elderly population, aiming to improve their quality of life and independence. These networks are composed by wireless sensor nodes that measure physiological variables and transmit it, for example, to the patients' smartphone, which may acts as a gateway to a remote medical assistance service. A key element of such sensor nodes is the wireless transceiver as it often dominates the overall power budget. Therefore, to provide a high degree of energy autonomy, wireless transceivers with an ultra low power consumption are needed. In this book, a transceiver architecture and implementation is presented, which targets such highly power constrained applications and employs the Bluetooth low energy standard.

At the architectural level, four main strategies are identified to obtain an ultra low power consumption. First, a direct-conversion receiver architecture is selected as it relaxes the requirements for the local oscillator and allows for a low power baseband section. Secondly, the number of active radio-frequency (RF) blocks has to be minimized in order to end up with as few RF nodes as possible that have to be driven by power-hungry circuits. Third, the remaining RF nodes have to be implemented with a high impedance level, which leads to a low required transconductance in the driving blocks and so reduces the power consumption. Finally, a low complexity demodulation scheme avoiding quadrature multi-bit analog-to-digital converters (ADCs) is needed.

The resulting transceiver architecture employs a passive receiver frontend architecture and a transformer at the antenna interface to boost the internal RF impedance. The carrier frequency is generated by a quadrature voltage controlled oscillator (QVCO), which is directly modulated to also synthesize the required signaling for transmission. In the baseband, the proposed transceiver employs a phase-domain ADC (Ph-ADC) which needs only 4 bits of resolution to demodulate the received signal.

In order to further improve the energy autonomy of the wireless sensor node, the possibilities for including an RF energy harvester with the presented transceiver frontend are studied. A fundamental problem that has to be resolved is the decoupling of the harvester from the transceiver while using the same antenna. In the proposed architecture the harvester is decoupled with an RF-switch that can be turned on passively, i.e., by utilizing the incoming RF power only. This approach stands out due to its low degradation of the transceiver performance as well as its small area occupation and hence low implementation cost.

On circuit level, the main contributions of this book are (a) a passive cancellation network to reduce the magnetic coupling-induced quadrature error of the QVCO, (b) a simple 4-transistor cell to directly modulate the QVCO tank capacitance in aFsteps, which has been verified to be sufficiently stable within the industrial temperature range, (c) a new current-domain linear combiner to provide the phase generated signals for the 4-bit Ph-ADC, which allows for a both area- and power-efficient implementation, and (d) an RF-switch that can be turned on without an external power supply using a start-up rectifier.

The proposed transceiver is implemented in a 130 nm CMOS technology using four integration steps which progressively complete the transceiver with the harvester. In the receive mode, the measured power consumption of 1.1 mW advances the state of the art as it is the lowest reported for a narrowband transceiver in the 2.4 GHz ISM band, which fulfills one of the typical WBAN standards. With a sensitivity of -81.4 dBm the receiver also achieves a competitive performance providing a sufficient link budget for a short-range data link. Concerning the performance in transmit mode, the power consumption of 5.9 mW and 2.9 mW in normal and back-off mode, respectively, is among the lowest reported so far for narrowband transmitters. However, the total transmitter efficiency of up to 24.5 % is significantly higher compared to other implementations due to the increased internal RF impedance. The harvester achieves a decent peak efficiency of 15.9 % and is able to progressively charge up an energy storage device for pulsed input signals emitted by an active WLAN router, for an expected distance of up to approximately 30 cm. Measurements also verify that the degradation of the transceiver, which arises from sharing the same antenna interface with the harvester, is less than 0.5 dB.

In conclusion, an ultra low power transceiver architecture for WBAN applications is presented which advances the state of the art in various aspects, as verified experimentally. Also the compatibility of the proposed architecture with energy harvesting techniques is shown, providing a possibility to improve the energy autonomy of wireless sensor nodes.

This book is organized as follows. After an introductory Chap. 1, Chap. 2 reviews the state-of-the-art of wireless low power transceivers. Then, Chap. 3 presents the four main strategies to reduce the power consumption and Chap. 4 describes the proposed transceiver in detail. Chapter 5 shows the co-integration of the energy harvester and finally Chap. 6 concludes this thesis.

Sevilla, July 2012

Jens Masuch
Manuel Delgado Restituto

Contents

Chapter 1
Introduction

The technological progress in wireless communication has greatly changed our habits in the past decades, mainly perceived today by the prominent role that smartphones play in our life. Especially, the miniaturization and cost reduction due to modern submicron CMOS processes has allowed to integrate a variety of services in such handheld devices. Following this trend, we can anticipate that smaller and smaller devices will be equipped with wireless interfaces in the future. Currently, peripheral wireless devices usually employ short-range communication protocols such as Bluetooth or ZigBee to form a wireless body area network (WBAN). However, since small devices are often highly constrainted with respect to the available power supply, the Bluetooth consortium has adopted in 2009 an extension to the standard, namely bluetooth low energy [3]. This new protocol is expected to greatly reduce the power consumption of the wireless interface and so promote many new applications.

Especially in the field of ambulatory health monitoring, WBANs have the potential to substantially improve the quality of life and independence of chronic patients and elderly people [116]. These networks are composed by wireless sensor or actuator nodes used for measuring physiological variables (e.g., glucose level in blood or body temperature [10]) or controlling therapeutic devices (e.g., implanted insulin pumps [86]). WBANs typically use a conventional star topology in which nodes acquire, process and transmit information to the central hub, that would be included in a smartphone for a typical ambulatory health monitoring application, as illustrated in Fig. 1.1. In this scenario, the smartphone can also serve as a real-time gateway to a medical control point or a remote assistance service by employing a long-range communication protocol [10, 86].

Wireless sensor nodes should exhibit a high degree of energy autonomy, which leads to the need for low-power consumption solutions in order to extend the battery lifetime or even make the node supply to rely on energy harvesting techniques. Typically, the power budget of the sensor node is dominated by the wireless link [48], and hence many efforts have been directed during the last years toward the implementation of power efficient transceivers (TRXs) [9, 14, 15, 18, 24, 26, 34, 36, 37, 41, 42, 45, 50, 57, 60, 62, 64, 67, 91, 101, 104, 109, 119, 120, 123, 133, 139,

J. Masuch and M. Delgado-Restituto, *Ultra Low Power Transceiver for Wireless Body Area Networks*, Analog Circuits and Signal Processing, DOI: 10.1007/978-3-319-00098-5_1,

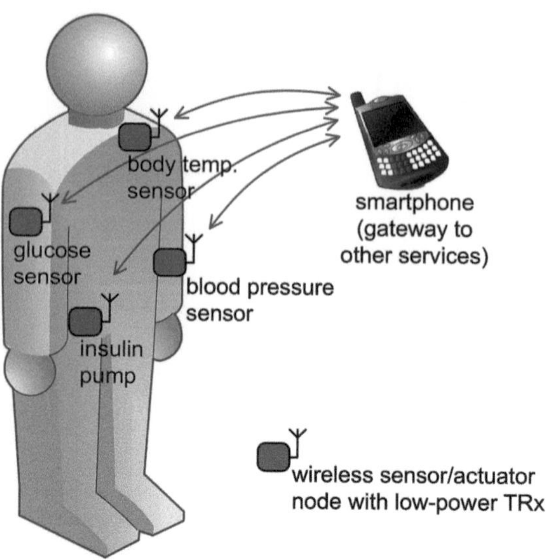

Fig. 1.1 Wireless body area network (WBAN)

140, 141, 147, 149, 151]. However, especially the most power efficient TRXs usually employ proprietary wireless interfaces which are often spectrally inefficient [14, 15, 18, 41, 45, 50, 91, 101, 120, 133, 139, 140, 149, 151]. Unlike to these solutions, the objective of this work is to reduce the power consumption while using a standardized and widely available wireless interface, namely bluetooth low energy (BLE). Sensor nodes that can easily connect to a smartphone without the need of proprietary read-out devices provide a second dimension of autonomy for the patient, apart from the battery lifetime. Additionally, the link layer of BLE supports AES encryption and key exchange algorithms to protect the highly sensitive personal data from unauthorized access. Finally, the 2.4 GHz industrial, scientific, and medical (ISM) band, in which BLE operates, is available worldwide and it allows for compact antennas designs, not larger than a few centimeters [2].

The methodology to reduce the power consumption employed in this book is rather an optimization at the architectural level than at the block level. For example, instead of optimizing a low-noise amplifier (LNA) for ultra low-power consumption, the necessity of an LNA at all is questioned. Considering the link budget needed for WBAN applications, an LNA is not necessarily required and a completely passive receiver (RX) frontend is sufficient. Similarly, questioning the necessity of an up-conversion mixer in the transmitter (TX) leads to the conclusion that the required signaling can be generated without a mixer by directly modulating the local oscillator (LO). Of course, optimization at the architectural level alone becomes worthless unless it is followed by a very power-aware design of the blocks.

The proposed transceiver has been implemented in standard CMOS technologies in order to allow for a low-cost solution, which is an important factor for wireless sensor nodes. At the beginning of this project, the feasibility of the quadrature LO was verified with two test chips (chip 1a and 1b) in a 90 nm 1P7M[1] CMOS technology. For availability reasons, the following implementations of the transceiver (chips 2–4) have been realized in a 130 nm 1P6M CMOS technology. Both technologies were provided by STMicroelectronics. The transceiver was implemented in two integration steps. In chip 2, the complete frequency synthesizer including PLL and direct modulation was integrated. To provide the realistic loading of the LO and to test the passive RX concept, also a provisional transceiver frontend was implemented in this run. In chip 3, the frontend was refined and a phase-domain demodulator was added to the RX baseband, completing the transceiver. In the final integration step (chip 4), an RF-to-DC converter was added to the frontend in order to show that the selected topology is also suitable for co-integration of RF energy harvesting without significant degradation.

1.1 Project Objectives and Organization

The main objective of this book is the implementation of a BLE transceiver in a low-cost CMOS technology with an ultra low-power consumption. Obviously, the BLE standard as well as the targeted application impose a set of specifications which have to be fulfilled, especially the blocking requirements with respect to near-by interferers. Complying with these requirements is particularly important considering that the transceiver operates in the 2.4 GHz ISM band, which is used by many applications. To provide a concise and handy overview, the specifications for the BLE signaling and the blocking requirements are summarized in the Appendix A.

A key requirement given by the application is the expected communication range. Taking into account channel imperfections, a certain link budget must be provided by the transceiver. Defining the link budget as the ratio of transmitter output power to receiver sensitivity (P_{out}/P_{sens}), a link budget of 80 dB is usually considered as sufficient to provide a robust communication link over a few meters. Therefore, the target output power of the transceiver is 0 dBm and the target sensitivity is −80 dBm.

Considering these performance requirements, the objectives with respect to the transceivers power consumption are defined. Currently, WBAN transceivers usually consume more than 10 mW and so dominate the available power budget. On the other hand, a power consumption on the order of 1 mW would be desirable, which is often referred to as the limit for autonomous operation in the literature, i.e., relying on energy harvesting only [115, 135, 143]. Given the fact, that the feasibility of sub-mW receiver frontends has been already demonstrated for proprietary 2.4 GHz transceivers [38], the target power consumption in RX-mode of the complete transceiver is set to about 1 mW. In TX-mode, the desired output power (0 dBm) sets the ultimate boundary of the power consumption. Considering that at least one-tenth of the dissipated power should be converted into RF output power delivered to the

---

[1] 1P7M means 1 Poly layer and 7 Metal layers.

antenna, a power consumption clearly below 10 mW in TX-mode is set as the design goal. Another requirement with respect to the power supply is to operate from the typical deep submicron CMOS supply voltage of 1.0 V in order to facilitate the integration of a sensor interface together with the transceiver in the future.

Finally, a supplemental objective of this project is to analyze the possibilities of including RF energy harvesting with as little impact on the transceiver as possible. This means that neither the TX output power nor the RX sensitivity shall be affected significantly while using the same antenna as the transceiver in order to avoid additional external components.

This book is organized as follows:

1. Chapter 2 provides a background about recent low-power transceivers for WBAN applications by reviewing the state of the art. In particular, the recent publications are categorized in three groups, namely narrow-band TRXs such as this work that usually comply with a WBAN standard, wide-band TRXs employing a proprietary signaling, and for completeness pulsed ultra wide-band TRXs.
2. Chapter 3 presents the architectural consideration for the proposed transceiver. Four main strategies are identified which are essential to achieve an ultra-low power consumption. First of all, the selection of the overall RX architecture is discussed with its impact on the individual blocks, which eventually results in a zero-IF architecture for the proposed transceiver. Second, the number of active RF blocks has to be minimized in order to end up with as few RF nodes as possible that have to be driven by power-hungry circuits. Third, the remaining RF nodes have to be implemented with a high impedance level, because this leads to a low-required transconductance in the driving blocks and so reduces the power consumption. Finally, a low-complexity demodulation scheme avoiding quadrature multi-bit analog-to-digital converters (ADCs) is needed. Therefore, the proposed transceiver employs a phase-domain ADC (Ph-ADC) which needs only four bits of resolution to demodulate the incoming signal.
3. Chapter 4 describes the implementation of the transceiver at the circuit level. Being a key building block in any narrow-band receiver, this section begins with the design of the frequency synthesizer, in particular the evaluation of the topologies for quadrature generation with low-power consumption. Then, the selected topology, namely a quadrature voltage controlled oscillator (QVCO), is refined to improve the accuracy and to implement direct modulation. The second section describes the design of the RF frontend, where the main focus is to maximize the internal RF impedance using a transformer. In the third section, the base-band part of the receiver is detailed, whose key element is the demodulator using a 4-bit Ph-ADC. Finally, the performance of the transceiver as a whole is evaluated, also in conjunction with a commercially available BLE transceiver. In TX-mode a power consumption of 5.9 mW is measured for delivering an output power of 1.6 dBm to the antenna, which corresponds to a total efficiency of 24.5 %. In RX-mode, the transceiver consumes only 1.1 mW and achieves a sensitivity of −81.4 dBm using a passive RF front-end topology without LNA.

4. In Chap. 5 the possibilities for including an RF energy harvester with the previously presented TRX front-end are studied. A fundamental problem that has to be resolved is the decoupling of the harvester from the TRX while using the same antenna. In the proposed architecture, the harvester is decoupled from the TX with an RF-switch that can be turned on passively, i.e., the incoming RF power is converted by a start-up rectifier in order to activate the switch. To decouple the harvester from the RX, the non-linear impedances of the harvester and mixer are exploited. The proposed harvester also comprises a supply management circuit to charge an external energy storage device. The correct functionality of the harvester is verified experimentally using pulsed RF signals as emitted from wireless local area network (WLAN) routers showing that a large holding capacitance can be charged for a distances of up to about 30 cm. The measurements also verify that the harvester hardly affects the TRX performance, i.e., the degradation is less than 0.5 dB.

5. Finally, Chap. 6 presents the conclusions of this book, highlighting the most important contributions. Moreover, the measured performance of the presented transceiver is compared to the state-of-the-art based on different performance metrics. Also, an outlook to possible future work is given.

Chapter 2
Review of the State of the Art

Given the wide range of possible applications, low-power short-range transceivers have drawn a lot of attention, both in research and industry. In this chapter, the current state-of-the-art-of transceivers operating in the frequency range from 1 to 10 GHz is reviewed briefly. Given the limited amount of data that is to be exchanged with wireless sensors or actuators, the WBAN transceivers usually implemented a data rate between 100 kb/s and a few Mb/s. Generally, the proposed transceivers can be categorized in three generic groups. First of all, there are conventional narrow-band transceivers which usually operate in the 2.4 GHz ISM band and implement one of the typical WBAN standards ZigBee [24, 60, 62, 64, 104, 109, 147], Bluetooth [26, 34, 36, 57, 67, 119, 123] or bluetooth low energy (BLE) [9, 37, 42, 141]. The second group are wide-band transceivers, which occupy a much larger bandwidth than absolutely necessary for their respective data rates in order to allow for super-regenerative receivers [18, 91, 101, 133, 149]. Finally, the third group are impulse-radio ultra wide-band (IR-UWB) transceivers that transmit extremely short RF pulses, and hence occupy a large bandwidth of several GHz [14, 15, 41, 45, 50, 120, 139, 140, 151].

2.1 Low-Power Narrow-Band Transceivers

Narrow-band transceivers are characterized mainly by their effective usage of the available spectrum, i.e., the signal bandwidth B is on the order of the data rate R. Therefore, the available spectrum can be split up into various channels and so allow for multiple users operating at the same time. This makes these transceivers attractive for commercial applications where inter operability and spectrum sharing are of particular importance. Consequently, the three dominating WBAN standards ZigBee, Bluetooth, and BLE define a narrow-band physical layer which exploit the 2.4 GHz ISM band (2.400–2.4835 GHz) with a different number of channels,[1] as illustrated in Fig. 2.1. Note that the ZigBee standard employs direct sequence spectrum spreading

---

[1] To be more precise, ZigBee employs the physical layer defined in the IEEE 802.15.4 standard.

J. Masuch and M. Delgado-Restituto, *Ultra Low Power Transceiver for Wireless Body Area Networks*, Analog Circuits and Signal Processing, DOI: 10.1007/978-3-319-00098-5_2,
© Springer International Publishing Switzerland 2013

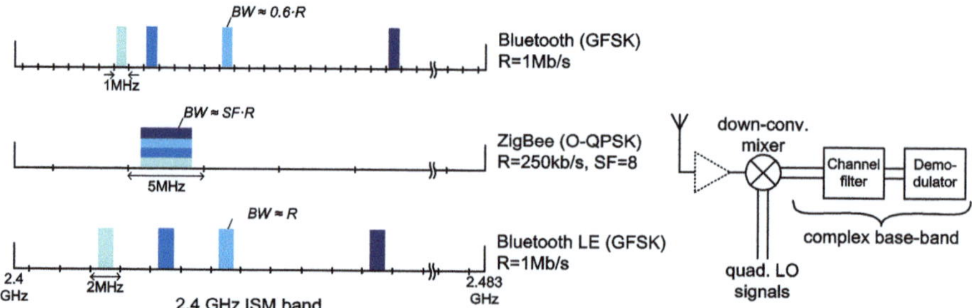

Fig. 2.1 Spectral planning of the typical narrow-band WBAN standards with the typical receiver architecture

(DSSS) to avoid interference, which effectively adds redundancy to the signaling, and hence increases the bandwidth. Increasing the redundancy by the ZigBee spreading factor of $SF = 8$ reduces the required signal-to-noise ratio at the receiver by 9 dB ($10 \cdot \log SF$,) and hence allows more users to operate in the same channel. Therefore, ZigBee transceivers are considered here as narrow-band systems, because the bandwidth B is on the order of $R \cdot SF$. Bluetooth and BLE avoid interferers by frequency hopping, which frequently changes the channel.

Narrow-band systems usually employ digital phase- or frequency modulation techniques such as Gaussian Frequency Shift Keying (GFSK) or Quadrature Phase Shift Keying (QPSK) for signaling. In fact, the Offset–QPSK scheme defined in ZigBee is equivalent to GFSK with a modulation index of $h = 0.5$ [60], and hence also similar to the modulation schemes employed by Bluetooth (GFSK, $h = 0.32$) and BLE (GFSK, $h = 0.5$). These modulation schemes have the advantage that they are spectrally efficient due to their low-modulation index and they provide a constant-envelope signal. The constant RF amplitude makes the system robust against nonlinear distortions and so allows for efficient power amplifiers in the transmitter.

To demodulate narrow-band signals under the presence of adjacent channel interferers, quadrature downconversion into a complex baseband is inevitable. The complex baseband allows for narrow channel filtering and hence for rejection of close-by interferers. Therefore, the essential building blocks of narrow-band receivers are the local oscillator (LO) defining the channel to be demodulated and a quadrature down-conversion mixer.

2.2 Super-Regenerative Wide-Band Transceivers

A lower power consumption can be achieved by using a higher bandwidth for the same data rate, i.e., $B \gg R$. This leads to a far less efficient usage of the radio spectrum, and therefore wide-band transceivers have been mainly proposed for closed systems in an academic environment [18, 91, 101, 133, 149]. These systems usually also employ FSK modulation but with a much larger modulation index ($h \gg 1$) [18, 101, 149], meaning that to transmit binary data symbols, two widely separated frequencies

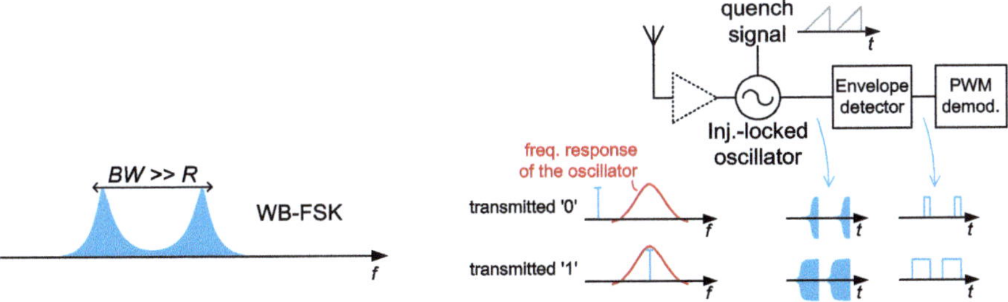

Fig. 2.2 Wide-band FSK (WB-FSK) power spectrum (*left*) and the typical super-regenerative receiver architecture

are transmitted, as shown in Fig. 2.2. The high modulation index allows for a simple, and hence power-efficient super-regenerative receiver architecture. However, this RX architecture is characterized by a poor spectral selectivity, which means that no interferers can be tolerated within a wide bandwidth. For this reason, also super-regenerative transceivers using On Off Keying (OOK) [91, 133] are considered here as wide-band systems.

Super-regenerative receivers employ an extremely simple architecture, which is based on injection locking of an oscillator. The input signal is applied to an oscillator whose resonance frequency is close to the expected incoming tone. Then, the start-up time of the oscillator is highly dependent on whether or not an incoming tone is present, i.e., a tone close to the resonance frequency stimulates the oscillator leading to a low start-up time, as illustrated in Fig. 2.2. By repeatedly ramping up the oscillator, which is referred to as *quenching*, the presence of the tone can be easily is monitored. Changing the resonance frequency oscillator also allows for monitoring tones at different frequencies and hence FSK demodulation, provided that the two FSK tones are sufficiently separated to make a difference in the start-up behavior of the oscillator. Therefore, the spectral selectivity of this demodulation scheme is mainly given by the bandpass characteristic of the oscillator, and hence depends eventually on the quality factor of the resonator. Given the low quality factor achievable in integrated solutions this leads to a poor spectral selectivity of the super-regenerative receiver architecture.

On the other hand, the simplicity of the super-regenerative RX architecture with extremely few blocks operating at RF allows for a very low-power consumption. Also on the transmitter side, both WB-FSK and OOK modulation allow for simple and low-power architectures, i.e., direct modulation of the oscillator or the power amplifier, respectively.

2.3 Impulse-Radio Ultra Wide-Band Transceivers

The third group is formed by impulse-radio ultra wide-band transceivers [14, 15, 41, 45, 50, 120, 139, 140, 151], which often achieve an even lower power consumption than the previous two groups. The DC power advantage of these transceivers can be

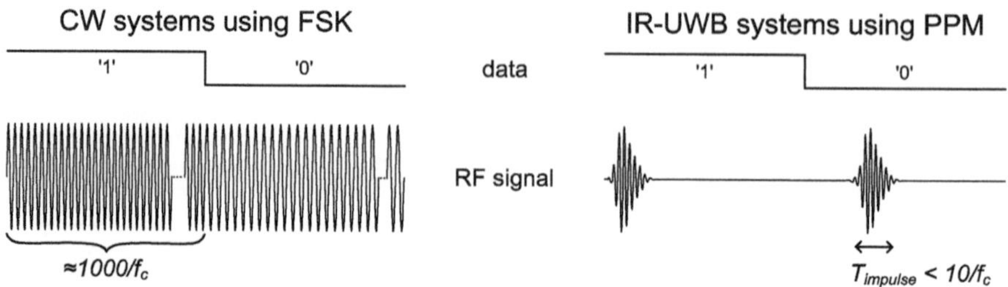

Fig. 2.3 Time-domain diagram of FSK-modulated CW systems (*left*) versus IR-UWB signaling (*right*) which transmits far less periods of the carrier (f_c) per data symbol

best observed in the time domain, as shown in Fig. 2.3. The first two groups essentially employ a continuous wave (CW) carrier signal (1–10 GHz) which is modulated with a low-frequency data signal ($\approx$1 MHz), and hence transmit thousands of RF periods per bit. In contrast to these CW systems, IR-UWB transceivers emit only short pulses per bit, usually comprised less than 10 RF periods. The data are often encoded with pulse position modulation (PPM) or with pulsed FSK. Therefore, the RF section of the transmitter is active only for a short fraction of the time, which leads to a low average power consumption. Once the transmitter and receiver are synchronized, the receiver can be duty cycled as well to reduce the power consumption for low data rates [45, 50].

The short RF pulses with an impulse duration (T_{impulse}) of usually less than 1 ns consequently occupy a very large bandwidth on the order of a few GHz ($\approx 1/T_{\mathrm{impulse}}$). This makes IR-UWB systems robust against narrow-band interferers and frequency-selective fading effects. On the other hand, the tolerance concerning other pulsed interferers depends highly on the synchronization capability. Moreover, the synchronization often dominates the overall RX energy dissipation, especially for short data packets [82, 140].

Although IR-UWB is a relatively new trend in WBAN communications, a standardized physical layer has been already defined (IEEE 802.15.4a) which operates in the frequency band from 3.1 to 10.6 GHz. Moreover, IR-UWB transceivers start to appear in the industry for point-to-point applications [139].

2.4 Comparison

To compare the three groups, Fig. 2.4 shows the sum of the TX and RX power consumption $P_{DC,TX+RX}$ of the relevant low-power transceivers versus their link data rate. The first group of narrow-band TRXs (shown in blue) can be clearly identified at the top with a total power consumption ranging from about 10 to 100 mW. Usually, one order of magnitude less power is consumed by the wide-band TRXs (1–10 mW), which also applies to the two ultra-low data rate TRXs with 5 kb/s not shown in the diagram [91, 101]. In contrast, pulsed TRXs are not distributed

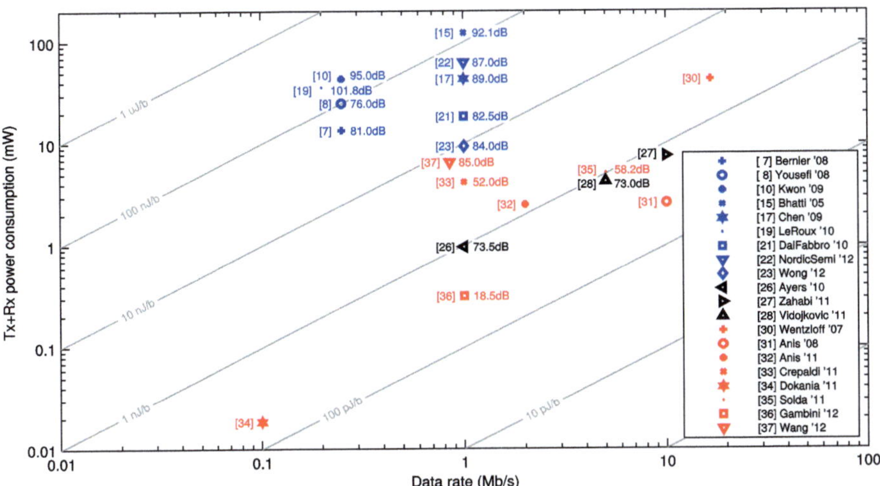

Fig. 2.4 Comparison of recent low-power transceivers considering the sum of TX and RX power consumption (y-axis). Narrow-band, wide-band, and IR-UWB transceivers are distinguished by *blue*, *black* and *red markers*, respectively. The dB-values in the figure denote the link budget of the respective transceiver (P_{out}/P_{sens})

within a characteristic DC power range but rather along a specific Energy-per-bit ratio on the order of 1 nJ/b, i.e., the power consumption of these duty-cycled TRXs scales with the data rate. However, around the typical WBAN data rate of 1 Mb/s the IR-UWB systems can achieve a power consumption of 1–2 order magnitude less than narrow-band TRXs [50].

Apart from the power consumption, an important figure of merit is the link budget (P_{out}/P_{sens}) provided by the transceivers. As mentioned before, a link budget of approximately 80 dB is needed for a robust short-range link [133, 139] taking into account small antennas and other channel imperfections such as fading. The annotated dB-values in Fig. 2.4 show that largest link budgets are obtained by narrow-band transceivers, hence promising the most robust operation. On the other hand, IR-UWB transceivers usually provide a much lower link budget of less than 60 dB due to the low average transmitted power. Although they are also less susceptible to fading effects, IR-UWB systems therefore usually offer a lower communication range than their narrow-band counterparts. Again, the wide-band TRXs fill the gap between the other two systems with link budgets around 70–75 dB.

To conclude, the IR-UWB systems that achieve the lowest power consumption either provide very low link budgets [50] or do not solve the synchronization issue [45]. Wide-band transceivers are promising with respect to power consumption and decent link budget but spectrally inefficient and prone to interference. The best interference tolerance, and hence inter-operability with other services can be achieved by narrow-band transceivers due to the complex baseband channel filtering. Moreover, such transceivers can connect easily to the existing handheld terminals as long as one of the typical WBAN standards is implemented. However, the total power consumption of such systems has to be further reduced in order to provide a higher degree of energy autonomy than the existing solutions.

Chapter 3
Low Power Strategies

3.1 Zero-IF RX Architecture

The definition of the receiver architecture is a fundamental design decision which drastically impacts the achievable power consumption. To select the appropriate receiver architecture, the requirements of the receiver in terms of spectral selectivity have to be taken into account. The Bluetooth low energy standard operates in the 2.4 GHz Industrial, Scientific, and Medical (ISM) band, which is shared with many other services. Therefore, in-band interferers have to be suppressed sufficiently to allow for correct demodulation of the GFSK input signal. In the BLE standard, interferers with an input power level 17 and 27 dB higher than the desired signal power must be tolerated at an offset frequency of 2 and 3 MHz or more, respectively [3]. Assuming a signal-to-noise ratio (SNR) of the demodulator of 15 dB, the interference suppression at these frequency offsets has to be at least 32 and 42 dB, respectively. This requirement already disqualifies the super-regenerative receiver architecture for the targeted application.

Narrow-band receivers for low power applications usually employ either a low intermediate frequency (IF) [57, 119] or a zero-IF down-conversion receiver [24, 109]. The low-IF architecture, shown in Fig. 3.1a, down converts the received signal to an IF which is usually on the order of the signal bandwidth, i.e., a few MHz. Then the adjacent channel interferers are filtered out by means of a complex band-pass filter, which not only removes the interferers at directly adjacent channels but also interference at the negative IF, also referred to as image frequency. The principal advantage of the low-IF architecture is that at no point signals around DC (0 Hz) are processed and so DC offset and flicker-noise problems are circumvented. On the other hand, to suppress the interferers at the image frequency accurate quadrature signals from the local oscillator are needed [48]. The achievable image rejection ratio (IRR) for a given quadrature accuracy can be calculated from

$$\frac{1}{\text{IRR}} = \frac{(\Delta A / A)^2 + \Delta \psi^2}{4} \tag{3.1}$$

J. Masuch and M. Delgado-Restituto, *Ultra Low Power Transceiver for Wireless Body Area Networks*, Analog Circuits and Signal Processing, DOI: 10.1007/978-3-319-00098-5_3,

where $\Delta A/A$ is the relative amplitude error and $\Delta\psi$ is the phase error of the quadrature LO signals [106]. Hence, to obtain an image rejection of 42 dB for example, only a phase error of $\Delta\psi = 0.9°$ can be tolerated if perfectly matched amplitudes are assumed. If amplitude errors were taken into account, the tolerable phase error would be even lower. To relax this stringent accuracy requirement for the LO, the BLE standard defines an exception of the interference rejection at the image frequency which permits interferers that are only 9 dB above the desired signal level instead of 27 dB [3]. This reduces the required IRR to 24 dB, which translates to a tolerable phase error of 7.2° (assuming equal amplitudes again).

An advantage of the zero-IF architecture is that it does not need the exception of the interference rejection, because the image-rejection issue is nonexistent. As shown in Fig. 3.1b, the desired signal is down converted to DC, and hence the interferers end up at higher baseband frequencies, where they can be eliminated easily using a low-pass filter (LPF). Regarding the accuracy of the LO signal, the zero-IF architecture is naturally robust against quadrature phase errors, i.e., a phase error of 5° degrades the receiver performance by only 1 dB [105]. On the other hand, the zero-IF architecture in principle suffers from DC offset and flicker noise which appear within the bandwidth of the desired signal. However, GFSK signals as specified in BLE can still be demodulated correctly even if the low-frequency contents below 19 kHz are removed [114], and hence these problems can be solved by a high pass filter (not shown in Fig. 3.1b).

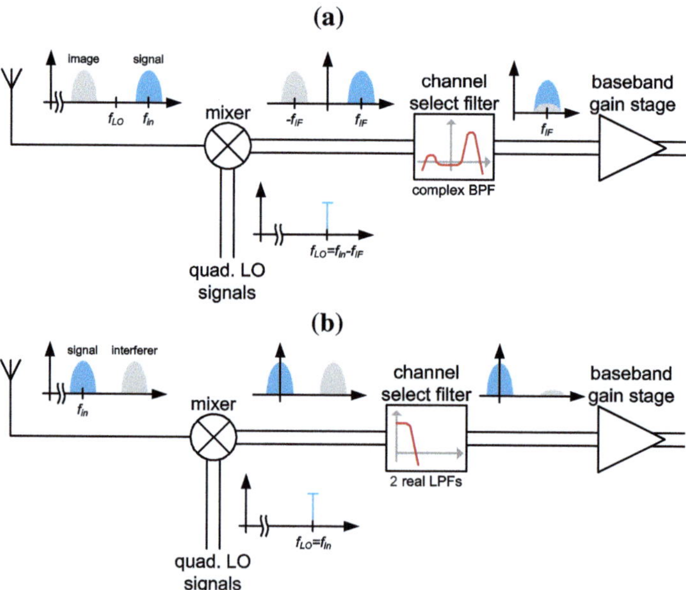

Fig. 3.1 Two typical narrow-band low-power receiver architectures: **a** low-IF architecture, **b** zero-IF architecture

Another advantage of the zero-IF architecture over low-IF is related to the required order of the filters. Lower order filters are needed with the zero-IF approach, because the transition from passband to stopband is spread over a wider relative frequency range. For example, if we consider the most demanding interference suppression of 32 dB at $f_{stop} = 2\,MHz$ offset from the desired signal and a maximum passband frequency of $f_{pass} = 500\,kHz$, the zero-IF architectures leads to a ratio of $f_{stop}/f_{pass,max} = 4$. In contrast, considering for example an IF of $f_{IF} = 2\,MHz$ the stopband and passband frequency are simply shifted, i.e., $f_{stop} = 4\,MHz$ and $f_{pass} = 2.5\,MHz$, and the ratio f_{stop}/f_{pass} drops below 2. Hence, the transition from passband to stopband in this low-IF example must be implemented in less than one octave, while the zero-IF approach leaves two octaves for this transition allowing for a smaller filter order. Finally, the zero-IF architecture also relaxes the gain-bandwidth-product requirements of the baseband section, because the signal is at lower frequencies and so favors more power-efficient implementations.

In conclusion, the zero-IF receiver architecture is selected for the low-power transceiver for the following reasons:

- the zero-IF architecture has no image-rejection issue and so needs no image exception in the interference blocking specification,
- it is robust against quadrature errors of the LO,
- it allows for simple, low-order LPFs for channel filtering, and
- DC offsets and flicker noise can be filtered with an HPF while still allowing GFSK demodulation.

3.2 Minimize the Number of Active RF Blocks

Although the super-regenerative RX architecture cannot be employed for a BLE transceiver due to the selectivity requirements, it is worthwhile analyzing why this architecture achieves such a low-power consumption and trying to apply similar concepts in a narrow-band RX. The key property of super-regenerative RXs is that the LO is used in an injection-locked manner, which directly provides a base-band output for demodulation. Low-power consumption is mainly caused by the fact that very few blocks are operating at the radio frequency, i.e., in the extreme case only the LO [17] which is often preceded by an amplifier to improve the sensitivity [19, 91, 101, 133]. Therefore, applying this concept to a narrow-band transceiver means minimization the number of active RF blocks. This requires to reassess the necessity of every RF block of the conventional narrow-band transceiver architecture instead of only optimizing each block.

If we consider the typical narrow-band RX architecture of Fig. 3.2a [57, 64, 109, 119], the good selectivity is achieved by performing the RX channel selection in the complex baseband after a quadrature down-conversion stage. Therefore, two ingredients are inevitable to this end, i.e., a quadrature down-conversion stage and quadrature local oscillator (LO) signals. However, the vast majority of low power

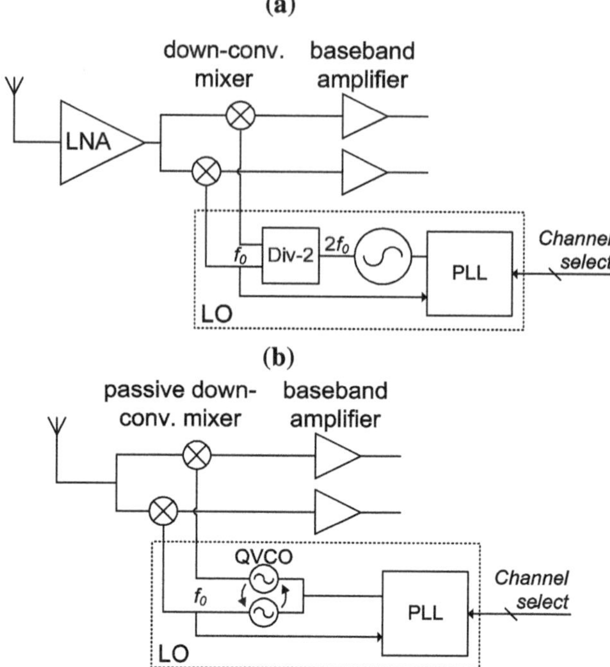

Fig. 3.2 Narrow-band single down-conversion receiver architectures: **a** typical topology with LNA and divider-based LO, **b** ultra-low power topology with a passive front end and a quadrature VCO (QVCO)

transceivers also employ a low noise amplifier (LNA) at the input of the RX chain. Note that even those LNAs which are optimized for low-power consumption usually consume 1–2 mW while achieving noise figures of 4–6 dB [23, 66, 131]. Hence, instead of trying to optimize the power consumption of the LNA for the given application, the question has to be answered if we need an LNA at all. In our case, for a short-range transceiver in a BAN application a sensitivity of $P_{\text{sens}} = -80\,\text{dBm}$ is sufficient to allow for a communication distance of 4 m [133]. Therefore, assuming again a moderate demodulator performance, i.e., a required SNR of 15 dB for GFSK demodulation [68], and a signal bandwidth of $B = 1\,\text{MHz}$ the tolerable noise figure NF_{max} can be calculated using

$$P_{\text{sens,dBm}} = -173.8\,\text{dBm/Hz} + 10\log BW + \text{NF}_{\text{max,dB}} + \text{SNR}_{\text{dB}} \qquad (3.2)$$

where $-173.8\,\text{dBm/Hz}$ denotes the fundamental thermal noise floor.[1] The resulting tolerable NF_{max} is as large as 18.8 dB and so allows to completely omit an LNA,

---

[1] The thermal noise floor is given by $k_B T$, where k_B is the Boltzmann constant ($k_B = 1.38 \cdot 10^{-23}\,\text{Ws/K}$) and a temperature of $T = 300\,\text{K}$ is assumed.

as shown in Fig. 3.2b [24, 38]. In addition, such a high NF_{max} allows to employ a passive mixer which obviously does not provide any gain, but also does not consume any DC power [38]. Hence, the only RF block consuming power in RX-mode is the LO which drives the passive mixer. The overall noise figure (NF) of this passive RX front end is mainly defined by the losses of the antenna interface and the conversion loss of the passive mixer, i.e., as in any RX front end the NF is limited by the losses until the first stage of active amplification, which in this case is the first baseband amplifier.

Note that some low-power TRX employ two down-conversion stages with a sliding IF [34, 62, 67, 141], meaning that both mixing signals are derived from the same LO through frequency division. However, although this slightly reduces the required LO frequency and allows for quadrature generation at a divided frequency, it excludes the passive front-end concept, because at least one buffering stage is needed between the mixers. In addition, the sliding-IF architecture would also require an additional up-conversion stage in the TX path, and is therefore discarded for the desired ultra low-power TRX architecture.

Regarding the quadrature LO for the single down-conversion architecture, low-power TRXs usually either employ one voltage-controlled oscillator (VCO) operating at twice the carrier frequency followed by a divider-by-2 [20, 57, 64, 109], as shown in Fig. 3.2a, or two VCOs which are coupled to each other to oscillate in quadrature [24, 29, 99, 104], as in Fig. 3.2b. The former alternative requires less chip area as only one VCO is needed and it is more robust against frequency pulling effects, because the VCO operates at a different frequency as the RX front end. However, the latter alternative with two VCOs, which is also referred to as quadrature VCO (QVCO), allows for a lower power consumption due to the lower oscillation frequency [11], which will also be demonstrated in a comparative study in the following chapter (Sect. 4.1.1). Hence, a passive receiver front end with a QVCO, as shown in Fig. 3.2b, is employed in the proposed low-power transceiver.

Using a QVCO also allows to directly drive the passive mixer without an intermediate buffer stage. Without a buffer the parasitic input capacitance of the passive mixer simply adds to the tank capacitance of the LC-VCOs and is therefore tuned out by the tank inductances. Note that LO buffers operating at 2.4 GHz usually consume at least several hundreds of μAs [20, 104, 146] and would consume a large portion of the targeted power budget in RX-mode of 1 mW. On the other hand, abstaining from buffering stages requires a careful co-design of all the front-end blocks in order to avoid undesired interactions. In order to avoid a substantial system degradation due to frequency pulling effects, a fast fractional-N phase locked loop (PLL) with a reference frequency of 20 MHz will be used. This allows for a high loop bandwidth of about 1 MHz in which pulling perturbations are attenuated by the PLL action [107]. Further, this high bandwidth leads to a short start-up time of about 5 μs for the synthesizer as well.

Taking a look at the TX section of the front end, different approaches can be used for GFSK modulation as required by the BLE standard. High accuracy can be achieved by using the mixer-based approach shown in Fig. 3.3a [37, 57, 142]. In this topology, a digital modulator processes the TX data to generate a set of quadrature

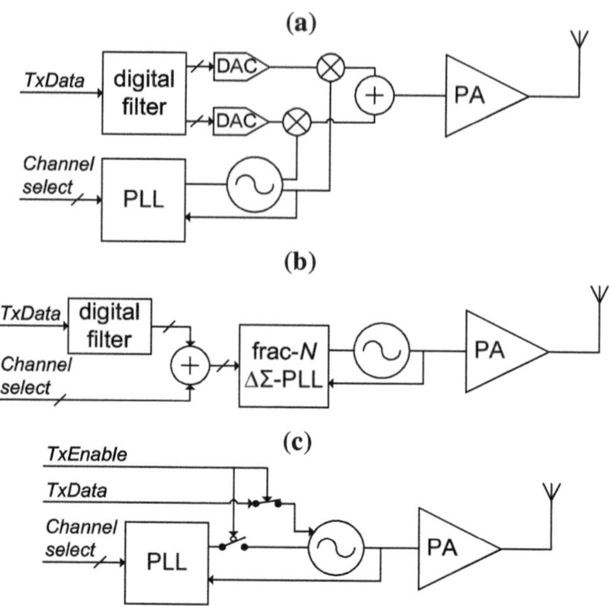

Fig. 3.3 TX architectures: **a** mixer based, **b** closed-loop modulation using a $\Delta\Sigma$-PLL, **c** open-loop using direct VCO modulation

baseband signals. These I- and Q-signals are then converted into the analog domain by means of digital-to-analog converters (DACs), and finally frequency shifted in one or more steps to the carrier frequency by means of mixers (in Fig. 3.3a a single-step direct up-conversion approach is represented). As modulation is realized in the digital domain, narrow-band channel filtering can be easily accomplished allowing for low transmission bandwidth occupation. Moreover, the architecture is highly flexible and can be used not only for FSK modulation but for any arbitrary modulation type. However, the power consumption of the mixer-based topology is high, typically above 10 mW [37, 57], which renders the approach poorly suitable for WBAN applications.

A second approach simply adds the FSK modulation to the carrier frequency setting of the phase-locked loop (PLL) as shown in Fig. 3.3b. Higher PLL resolutions are achieved by using fractional-N PLLs with $\Delta\Sigma$-modulators [92, 98]. In principle, this technique is both simple and accurate, because the FSK modulation is added in the digital domain. However, the loop bandwidth of the PLL acts as a low-pass filter on both the FSK modulation and the quantization noise of the $\Delta\Sigma$-modulator leading to a fundamental trade-off between transmission rate and phase noise. On the one hand, a low loop bandwidth is needed to filter out the quantization noise and, on the other, the loop bandwidth must be large enough to preserve the FSK modulation. In order to relax this trade-off, a possible solution consists on pre-emphasizing the FSK modulation signal to make its bandwidth much larger than that of the PLL [98]. Unfortunately, pre-emphasis can only be as accurate as the prediction of the PLL transfer function, and therefore it is hardly feasible in a fully integrated solution where the VCO gain is essentially nonlinear [12]. Alternatively, if a high PLL bandwidth is used, complex compensation techniques must be employed to adaptively cancel the quantization noise of the $\Delta\Sigma$-modulator. Such compensation schemes usually employ power-hungry DACs and raise the power consumption of the FSK modulation stage by several mWs [51, 92, 124].

Finally, the FSK modulated signal can also be generated by means of direct VCO modulation [34, 52, 101]. A PLL is used to set the carrier frequency before it is disconnected, so that modulation can be directly fed into the VCO, as shown in Fig. 3.3c. The main advantage of this technique is that very little additional hardware is needed, thus leading to very low-power consumption. Moreover, the transmission rate is not constrained by the PLL bandwidth, which allows to increase the data rate easily up to several Mb/s [19]. However, as the PLL remains open during transmission, the unlocked VCO becomes susceptible to frequency drift due to leakage currents. Therefore, unlike the previous approaches, the technique is unsuitable for continuous modulation,[2] but it can be exploited in communication schemes that employ short transmission bursts such as the BLE protocol.

Therefore, the proposed BLE transmitter is based on the direct QVCO modulation technique. The BLE data packages have a maximum length of $376\,\mu s$, which is short enough to keep frequency drifts during data transmission within the specifications of the standard. Also, the permitted spread of the FSK modulation index of $h = 0.5 \pm 10\%$ is wide enough, so that direct VCO modulation, implemented in analog domain, is able to underlie process, supply voltage and temperature (PVT) variations.

In conclusion, the proposed low-power narrow-band transceiver architecture comprising a passive RX front end (Fig. 3.2b) and a TX with direct VCO modulation (Fig. 3.3c) is characterized by a rigorous simplification in terms of the number of RF blocks. Essentially, the architecture is reduced to its bare minimum, where the only active RF block in RX-mode is the LO and the down conversion is performed by passive mixers. A second important aspect is that buffering stages are avoided as much as possible which requires a careful co-design of the remaining blocks. This means that especially the LO has to be simulated always with all its loading blocks, i.e., the mixer, the PA and the prescaler, which forms the input stage of the PLL.

3.3 Maximize Impedance of Internal RF Nodes

The simplified front-end architecture of the TRX contains only three critical RF nodes that have to be driven by active circuitry. Therefore, these nodes have a great impact on the overall power consumption. The first two are the two LC-tanks of

---

[2] It should be noted that direct VCO modulation may also be applied on a low-bandwidth closed PLL if the data signal has no DC content [35]. In this case, the PLL is active all the time avoiding drift issues and only acts as a high-pass filter on the data signal without corrupting it. However, this technique is not applicable for the BLE standard with its DC-carrying line code.

the QVCO, while the third critical RF node is the output of the PA, i.e., the antenna interface.

The LO employs two cross-coupled LC-VCOs which are coupled to each other to operated in quadrature. In spite of the technology down scaling, such LC-oscillators [59, 81, 95] still consume less power than ring oscillators [28, 76, 96] at oscillation frequencies in the GHz range. As will be shown in detail in the next chapter (Sect. 4.1.1), the power consumption of LC-VCOs is ultimately limited the inductance that can be implemented in the tank. To reduce the required bias current, the parallel tank impedance has to be as large as possible. This means that large inductors with many turns have to be used which are operated close to their self-resonating frequency [129].

A similar relation exists for the antenna node, which is driven by the power amplifier. For the targeted output power of 1 mW a PA load impedance on the order of a few kΩs is needed to allow for a power efficient implementation [38, 111]. Since typical WBAN antenna impedances are on the order of 50–100 Ω [2], impedance up conversion is needed at the antenna interface. In the proposed front end, a step-up transformer is used to up convert the antenna impedance to about 1 kΩ. The transformer arrangement has been chosen, because it allows to connect a single ended or differential antennas [57]. Similarly as for the VCO nodes, this up-conversion requires to use a transformer with a high inductance at the PA coil in order to increase the parallel loss impedance of transformer, and hence make it non-dominant with respect to the up-converted antenna impedance of 1 kΩ. This means again, that a multi-turn transformer is needed that has to be operated in proximity of its self-resonance frequency.

Therefore, the impedance of both actively driven RF nodes is limited by the inductive elements which have to be chosen carefully. Since the inductive elements are also operated close to their self-resonating frequencies, the devices have to be well characterized with all its parasitic elements using three-dimensional electro-magnetic simulations.

3.4 Low-Complexity Phase-Domain GFSK Demodulator

As explained in Sect. 3.1, the zero-IF architecture is particularly suitable for ultra low-power transceivers, and therefore calls for zero-IF GFSK demodulators. In a straightforward approach, the quadrature input signals are quantized by two analog-to-digital converters (ADCs) and the channel filtering and demodulation tasks are performed in digital domain [32, 117]. However, the need for two power-hungry multi-bit ADCs disqualifies this concept for low-power applications. Alternative strategies for low-power GFSK demodulation can be conceived by taking advantage of the fact that information is carried in the signal phase alone, a property that is usually also exploited by limiting low-IF demodulators [27, 46, 57, 70, 119]. Ideally, the transmitted data are coded by the rotations of a constant magnitude phasor around the origin of the complex plane. A zero is coded as a clockwise rotation,

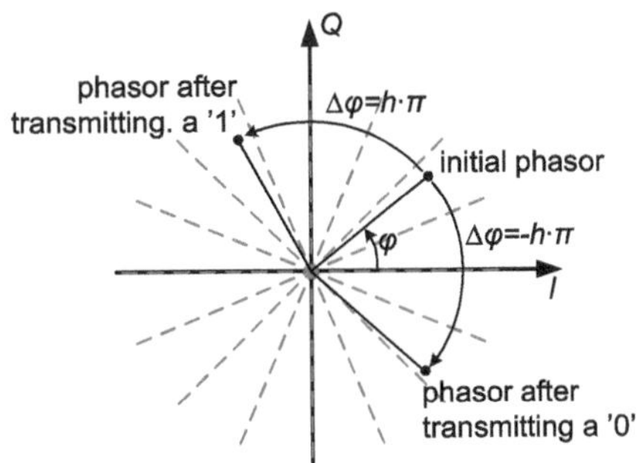

Fig. 3.4 (G)FSK modulation with a modulation index h in the complex plane. The *dashed lines* represent the levels of a 4-bit phase quantization

whereas a one leads to a counter-clockwise rotation, as illustrated in Fig. 3.4. Using this property, Lee and Kwon introduced the concept of direct phase quantization in their zero-IF zero-crossing demodulator (ZIFZCD) [68]. The ZIFZCD generates rotated versions of the in-phase and quadrature signal (I- and Q-signal) and detects the time and direction of their zero crossings [44, 68]. Then, the detection of the transmitted symbol is simply accomplished by comparing the number of clockwise and counter-clockwise crossings. This concept was further developed by Samadian et al. [114] where the outputs of the zero-crossing comparators are employed to obtain a phase read-out, which is nothing else than a phase-domain ADC (Ph-ADC).

The approach of phase quantization has two important advantages with respect to the conventional amplitude quantization of the two baseband channels. First of all, the Ph-ADC requires only 16 quantization intervals (4 bits) to demodulate GFSK with a modulation index of $h = 0.5$ while the conventional approach usually requires two separate ADCs with at least 6-bit resolution [32, 89, 117]. Hence, the amplitude quantization approach needs much more comparators assuming a flash-ADC architecture or a higher processing rate if sigma–delta or successive approximation architectures are used. In either case, the power consumption will be larger for the two-channel amplitude quantization.

The important second advantage of the phase quantization is its robustness against distortion [22] and its high input dynamic range, because only the phase information is processed. This renders an accurate amplitude equalization prior to quantization unnecessary, i.e., the gain equalization in the baseband can be performed in coarse steps which relaxes the requirements of the analog pre-processing. To obtain an equivalent dynamic range of a 4-bit Ph-ADC, two conventional ADCs with at least 10-bit resolution are required [22, 48].

Although the Ph-ADC demodulation concept is promising in terms of power consumption and bit-error-rate (BER) performance [114], few silicon realizations have been reported so far [21, 37, 67]. All these examples employ an array of resistors to generate the rotated I- and Q-signals. In the proposed transceiver, an alternative scheme for phase rotation is employed which does not require resistors but combines the weighted outputs of current mirrors instead, as will be detailed in Sect. 4.3 of the following chapter. This solution reduces not only the amplitude error of the phase rotation but also allows for an area-efficient implementation.

Chapter 4
Implementation of the Low Power Transceiver

This chapter describes the circuit-level details of the proposed low power transceiver, shown in Fig. 4.1. It is organized in three parts corresponding to the main functional units, namely the frequency synthesizer, the RF frontend, and finally the baseband section of the receiver.

4.1 Frequency Synthesizer

The main function of the frequency synthesizer is obviously to provide the carrier frequency for the 40 BLE channels. Therefore, the accuracy of this LO signal defines the spectral efficiency of the transmitter as well as the spectral selectivity of the receiver. However, the LO accuracy requirements for WBAN transceivers are usually not very demanding due to the low TX output power and the moderate selectivity requirements of such transceivers. In case of BLE, the blocking requirements impose a phase noise level below -92 or $-102\,\text{dBc/Hz}$ at an offset frequency of 1.5 or 2.5 MHz, respectively, assuming a signal-to-noise ratio (SNR) of 15 dB. Such phase noise requirements are easily achievable by a fully integrated synthesizer. Assuming again an SNR of 15 dB, the tolerable spur level is $-32\,\text{dBc}$ and $-42\,\text{dBc}$ at an offset of 2 and 3 MHz or more, respectively.

The receiver also requires quadrature LO signals, i.e., an in-phase and a quadrature component, to demodulate complex modulated signals such as PSK or narrow-band FSK, which is the case for BLE. Regarding the quadrature accuracy, the selected zero-IF receiver architecture helps to relax the requirements. As long as the quadrature error is below $5°$, the sensitivity is degraded by less than 1 dB [105]. On the other hand, the power consumption of the synthesizer plays a crucial role in the overall power budget of the transceiver because it is the only active RF block in RX-mode due to the passive frontend architecture. Therefore, it is important to choose the quadrature LO generation topology with minimum power consumption and sufficient accuracy. This selection process is detailed in the following subsection.

J. Masuch and M. Delgado-Restituto, *Ultra Low Power Transceiver for Wireless Body Area Networks*, Analog Circuits and Signal Processing, DOI: 10.1007/978-3-319-00098-5_4,
© Springer International Publishing Switzerland 2013

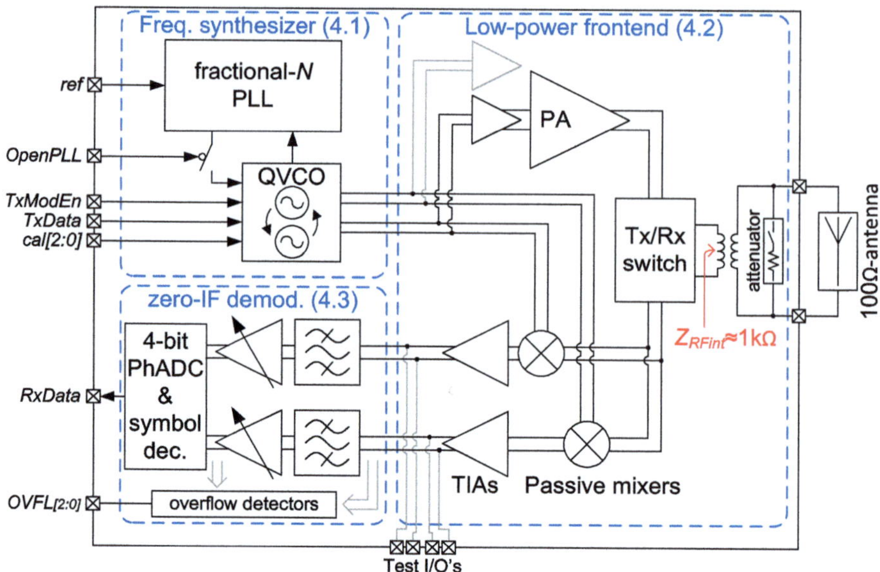

Fig. 4.1 Block diagram of the proposed BLE transceiver

4.1.1 Selection of Quadrature Generation Topology

Different alternatives have been proposed so far for the implementation of quadrature generators. Taking advantage of the increased speed of deep submicron CMOS technologies and the moderate phase noise requirements of WBAN communication standards, very simple and area-efficient ring oscillators can be used to directly generate quadrature outputs in the GHz range [28]. However, such ring oscillator topologies still consume more than 10 mW and are therefore not suitable for low power applications. Much more power efficient architectures can be implemented using the well-known cross-coupled LC VCO, which conserves part of the oscillation energy in its reactive components [90]. Using a single LC oscillator core running at the desired output frequency, quadrature phases can be obtained by means of RC–CR networks [90] or by cascading poly-phase filters [16, 60] at the output of the VCO. However, these passive structures always exhibit resistive losses or even lead to amplitude errors [90] which have to be corrected using power-hungry limiting amplifiers. To avoid these shortcomings, a promising solution consists of a VCO running at twice the desired output frequency and a frequency divider-by-2 block to generate the four quadrature output phases [20, 57, 64, 72, 95, 109, 121]. Another alternative relies on properly coupling two VCO cores at the target output frequency to generate the quadrature phases [11, 24, 29, 99, 103, 104, 130]. In this section, the suitability of the two last quadrature generation architectures for WBAN applications is explored and validated with corresponding silicon integrations in a 90 nm CMOS technology.

The first alternative, formed by the combination of a VCO and a frequency divider, hereinafter referred to as VCODIV2, is commonly used for its simplicity and area efficiency. Both building blocks can be connected in cascade [20, 57, 64, 109, 121] or stacked onto each other in order to reuse the bias current [72, 95]. The latter requires either a high supply voltage of at least 1.5 V [95] or a transformer which leads to an overall power consumption greater than 5 mW [72]. Accordingly, focus will be paid in the cascade solution, more suitable for operation in the 1 V supply range.

The second architecture is a quadrature structure, hereinafter referred to as QVCO, which uses two LC tanks that resonate at the target output frequency [11, 24, 29, 99, 103, 104, 130]. Such structures differ in how the two oscillator cores are coupled to each other, i.e., parallel or serial coupling. To minimize the required transconductance to start oscillation, the classical parallel coupling has been preferred. Again, it is also possible to stack the two cores and exploit current-reuse techniques [59]. However, since one core uses PMOS transistors while the other is based on NMOS transistors, the structure is asymmetric and, hence, prone to phase errors. In the proposed QVCO, each core employs current-reuse separately and, additionally, an arrangement of RC phase shifters is used in the coupling path between the two oscillating cores to reduce phase noise without any power consumption penalty.

4.1.1.1 VCODIV2 Architecture

In this architecture, an LC VCO running at about 4.9 GHz is followed by a divider-by-2 block to generate the four quadrature phases in the 2.4 GHz band. The implemented circuit is displayed in Fig. 4.2. The VCO core is a cross-coupled LC oscillator in which a simple MOS transistor arrangement provides the negative transconductance needed to compensate for the losses of a parallel LC tank. Such arrangement consists of complementary NMOS and PMOS sections that reuse the VCO tail cur-

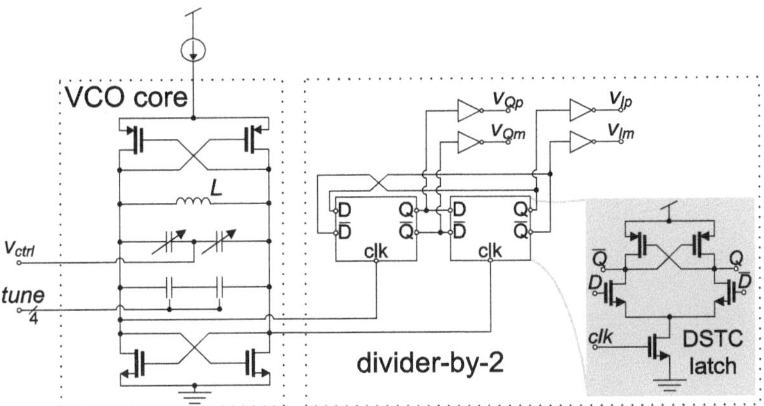

Fig. 4.2 Schematic of the implemented VCO with divider-by-2

rent [129]. This provides an extra mechanism for power saving as compared to single section realizations.

In order to minimize the required negative transconductance and, thereafter, the power consumption of the oscillator, the losses of the LC tank must be reduced as much as possible. In practice, the tank losses, which can be lumped into a parallel conductance g_{loss}, are dominated by the integrated inductor [90] and can be approximated as

$$g_{loss} \approx g_{ind} = \frac{1}{Q_L \omega_0 L} \qquad (4.1)$$

where g_{ind} is the loss conductance of the inductor, $\omega_0 = 2\pi f_0 = 1/\sqrt{LC}$, L is the inductance, and Q_L is the loaded quality factor of the inductor at tanks resonance frequency f_0. This frequency f_0 is made tunable by electrically controlling the value of the total tank capacitance C_{tank}.

To select the appropriate tank inductor, the equivalent models of the available differential inductors in the given technology are extracted from the design kit. These models contain frequency-dependent series resistances to represent the effect of eddy currents as well as substrate effects (parasitic substrate resistance and capacitance). Based on this data, the important design parameters, i.e., the loaded quality factor and the loss conductance, are calculated and shown in Fig. 4.3. Each curve in this diagram corresponds to a certain number of turns of the differential inductor. Note that candidate inductances are ultimately limited by the capability to tune the tank to the desired frequency. In Fig. 4.3, a minimum safe value for the tuning capacitance of 200 fF has been assumed. Denoting as C_{ind} the parasitic differential capacitance of the inductor and letting C_{tune} represent the remaining tank capacitance including parasitics, the aforementioned tunability condition can be expressed as

$$\frac{1}{\omega_0^2 L} - C_{ind} = C_{tune} > 200\,\text{fF}. \qquad (4.2)$$

The rightmost endpoints of the curves for 4 and 5 turns correspond to the situation where the tunability condition is just fulfilled.

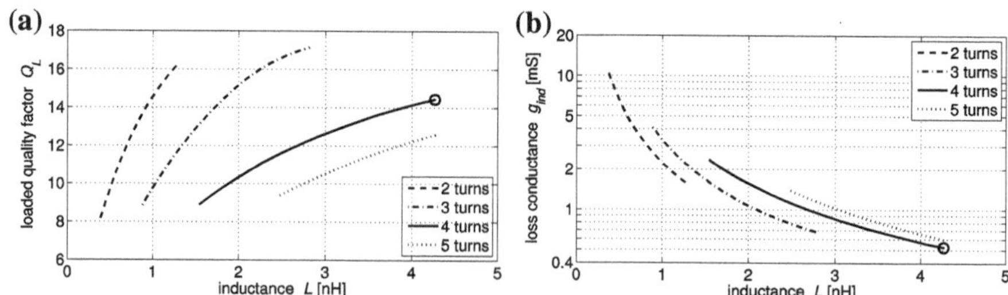

Fig. 4.3 Properties of the available symmetrical inductors at 4.9 GHz, the *circle* marks the selected inductor for the VCODIV2. **a** Loaded quality factor. **b** Loss conductance

Figure 4.3a shows that the maximum Q_L is reached at an inductance of approximately 2.8 nH obtained with a 3-turn spiral inductor. However, as shown in Fig. 4.3b, the loss conductance decreases for larger inductances. This implies a tradeoff between performance (phase noise improves with higher quality factors [90, 129]) and power consumption. Given the mild requirements in phase noise, g_{ind} reduction has been prioritized and, thereafter, the largest possible inductor has been selected. It is a 4.23 nH 4-turn inductor which achieves a loaded quality factor of $Q_L = 14.4$ @ 4.9 GHz. As it will be shown with the experimental results (Sect. 4.1.1.3), this value of Q_L is high enough to fulfill the phase noise requirements of the quadrature generator. Note, that this selection procedure eventually leads to an inductor that just meets the tunability condition (4.2). Therefore, the parasitic tank capacitance has to be minimized as much as possible during circuit and layout design.

Frequency tuning has been implemented using a conventional two-step approach. Coarse tuning to compensate for process variations is achieved with a programmable MOM capacitor[1] bank controlled by a 4-bit tune word. The coarse tuning range has been designed such that it exceeds the frequency error range due to process, supply voltage, and temperature (PVT) variations, which cause a resonance frequency deviation of about $\pm 5\%$ according to corner simulations (corners of the process, supply voltage $\pm 10\%$, and temperature range from -40 to 85 °C). PMOS varactors allow for fine frequency control by means of a phase-locked loop (not shown in Fig. 4.2). The simulated quality factor of the switched MOM-capacitor bank is larger than 60 over the digital tuning range whereas the quality factor of the PMOS varactor is larger than 65 over the whole analog tuning range. Therefore, the capacitive tuning of the tank has very little effect on the overall tank losses.

The active circuitry of the VCO core has been sized considering the difference between transconductance g_m and output conductance g_{ds} because they appear in parallel to each other in the small signal model of the core. This effective transconductance $g_m - g_{ds}$ has been maximized while keeping the total tuning capacitance to the targeted 200 fF. This gives 14/0.15 μm for the NMOS transistors and 14/0.12 μm for the PMOS transistors.

The divider-by-2 circuit is implemented with two latches forming a master-slave flip-flop. Dynamic single-transistor-clocked (DSTC) latches have been used instead of the more often used source-coupled logic (CML) latches for power saving [43, 148]. Taking into account the low amplitude of the oscillations generated by the low-power VCO (single-ended amplitudes of about 150 mV), the grounded-source input transistors of the divider can be regarded as current sources injecting an AC current into a differential ring oscillator [25, 84]. This view justifies that no buffering stage is used between the VCO and the divider. As the input transistors of the divider simply mirror the currents of the VCO into the ring oscillator, any common-mode compensation for PVT variations is unnecessary. The divider provides symmetric output phases due to its fully differential operation. Four CMOS inverters act as output buffers in order to make the divider immune to loading effects.

---

[1] Metal-Oxide-Metal capacitors are formed by interdigitated metal lines using the thin metal layers.

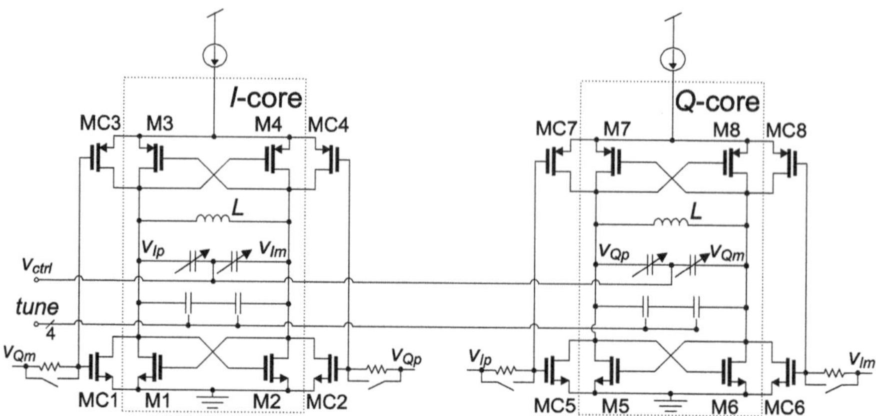

Fig. 4.4 Schematic of the implemented quadrature VCO

4.1.1.2 QVCO Architecture

The QVCO consists of two LC oscillator cores with equal tank properties, which are forced to operate in quadrature by means of a coupling mechanism, as shown in Fig. 4.4. The oscillation frequency of each of the cores is now equal to the desired output frequency of 2.45 GHz, i.e., half the frequency of the VCO core described before. Following a procedure similar as for the VCODIV2, the important inductor parameters at 2.45 GHz have been extracted from the design kit, as shown in Fig. 4.5. Again, the same tuning condition has been applied as before, i.e., leaving a 200 fF tank capacitance margin for the remainder of the tank. For the QVCO, this procedure leads to a 7-turn 13.6 nH inductor with a loaded quality factor of 11.5 @ 2.45 GHz. Note that this inductance is not four times larger than the one used in the previous architecture as it could be expected from halving the resonant frequency. This is because a larger inductor has a significantly higher parasitic capacitance C_{Ind} and hence the allowable range of inductances is reduced, according to the tunability condition (4.2). Again, coarse and fine tuning of the oscillation frequency has been

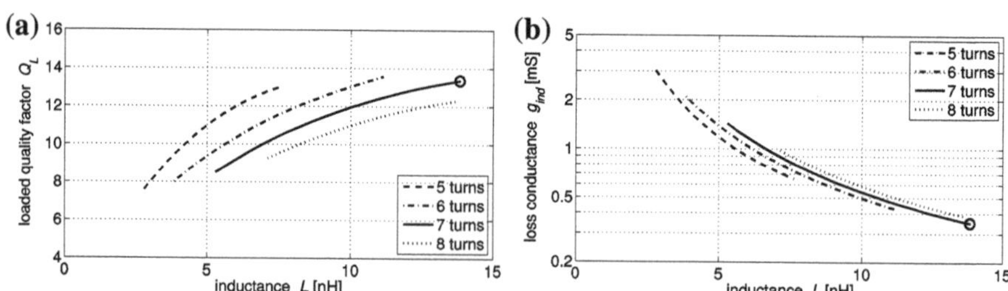

Fig. 4.5 Properties of the available symmetrical inductors at 2.45 GHz, the *circle* marks the selected inductor for the QVCO. **a** Loaded quality factor. **b** Loss conductance

implemented by a MOM capacitor bank and PMOS varactors, respectively. The same MOM capacitor bank used in the VCODIV2 design was found sufficient to cover the PVT variations. However, the size of the PMOS varactors had to be increased by 25 % in order to maintain the same tuning range and VCO gain, because of the larger tank capacitance of the QVCO. Due to the lower oscillating frequency, the simulated quality factors of the capacitive tuning elements are even higher than for the VCODIV2, namely, larger than 100 for the MOM capacitor bank and larger than 65 for the PMOS varactors. The same procedure used in the VCODIV2 prototype has been employed for dimensioning the active parts of the two LC oscillator cores leading to the similar transistor sizes, i.e., 12/0.15 and 12/0.12 μm for the NMOS and PMOS core transistors, respectively.

Coupling between the two cores of the QVCO can be accomplished through the addition of transistors in parallel or in series to the cross-coupled pairs. These coupling transistors combine a direct connection and a cross (inverting) connection between the two VCOs which force them to oscillate in quadrature [113]. Although coupling transistors in series [11] achieve a better phase noise performance, they increase the required supply voltage headroom and also the required bias current to start the oscillation. Therefore, a coupling mechanism based on parallel transistors, as shown in Fig. 4.4, has been chosen. In order to maintain the power consumption low, coupling transistors MCx have a width 4 times smaller than the cross-coupled pairs Mx.

In the conventional QVCO, the coupling transistors of the I-core are directly driven by the Q-core signals and vice versa [113]. Therefore, the coupling transistors operate at 90° phase difference with respect to their corresponding VCO core transistors. It has been shown, however, that reducing this phase difference toward ideally 0° improves the phase noise performance of the QVCO [85]. To achieve this goal, active phase shifting techniques can be applied at the price of increasing the power consumption [132]. In order to avoid any extra power dissipation, a fully passive solution has been applied in this QVCO design. It consists on using RC phase shifters formed by series resistances and the gate capacitances of transistors MCx, as shown in Fig. 4.4. This solution reduces the phase difference to approximately 45° but still noticeably improves the phase noise performance of the QVCO. In order to allow measurements with and without the RC phase shifters, the resistances can be bypassed using NMOS switches. As will be shown, in the following subsection detailing the experimental results, the proposed passive solution is able to improve the phase noise performance of the QVCO by 2 dB. Another advantage of using phase shifters in the coupling path is that the QVCO has only one stable mode of operation, i.e., the in-phase oscillation lags the quadrature phase by 90° [85].

4.1.1.3 Experimental Results

Both quadrature generation architectures have been implemented in a 90 nm CMOS process with 7 metal layers including 2 thick top layers (Fig. 4.6). The two different quadrature generators are identically loaded by output buffers (implemented by

(a)   (b)

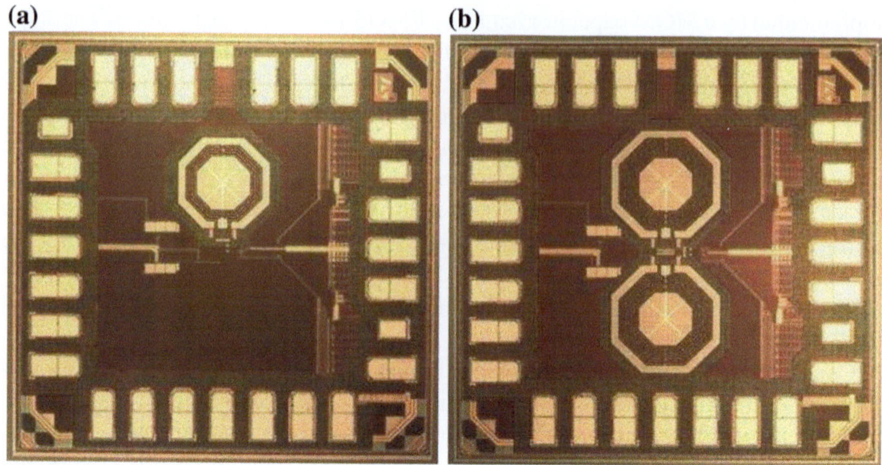

Fig. 4.6 Die photo of the test chips 1a and 1b (both die sizes are 1-by-1 mm). **a** VCODIV2, **b** QVCO

(a)   (b)

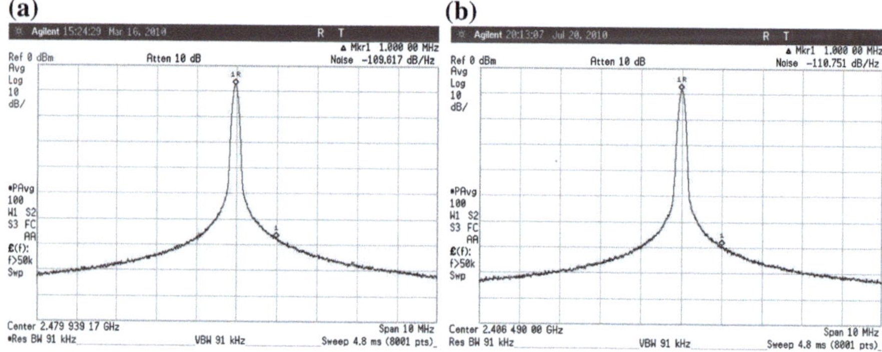

Fig. 4.7 Measured output spectrum of the quadrature generators. **a** VCODIV2, **b** QVCO

push-pull inverting amplifiers) and a phase-switching prescaler. They employ an identical pad ring and have been mounted in QFN-28 packages. Therefore, the same test board has been used to measure the chips. All the RF measurements have been performed using an Agilent E4440A PSA spectrum analyzer. The supply voltage of the chips has been set to 1 V. Output buffers have been supplied via a separate pin in order to allow for measurements of the current consumed by the generator cores.

Figure 4.7 shows an example of the measured output spectrum obtained with the two integrated quadrature generators highlighting the phase noise at 1 MHz offset. To allow for comparison with simulation results, the recorded output spectra have been normalized to the center frequency and scaled to a resolution bandwidth of 1 Hz. Figure 4.8 shows that the measured phase noise of the VCODIV2 circuit is in very good agreement with the simulation. The larger difference between simulation

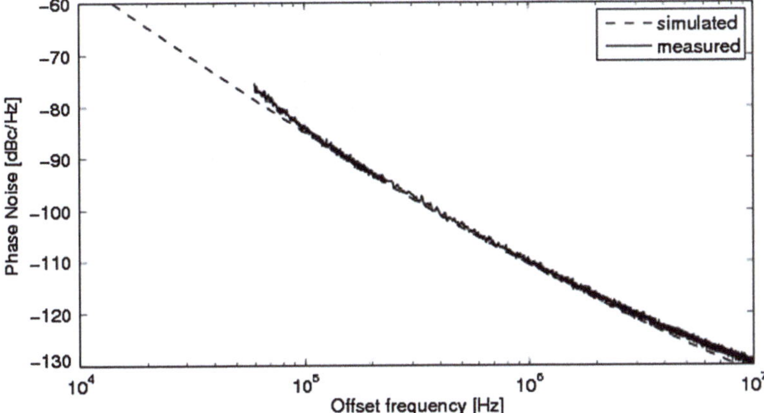

Fig. 4.8 Measured phase noise performance of the VCODIV2 circuit (*upper* sideband)

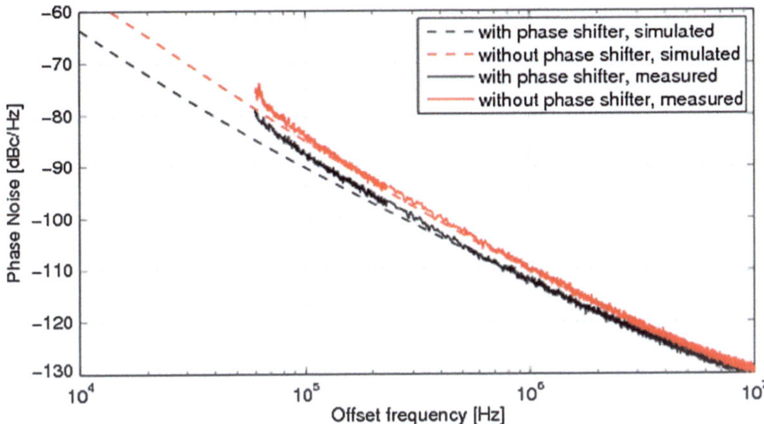

Fig. 4.9 Measured phase noise performance of the QVCO (*upper* sideband)

and measurement at lower offset frequencies indicates that the $1/f$-noise corner frequency is slightly higher than expected.

The measured upper sideband phase noise of the QVCO is shown in Fig. 4.9 for both modes of coupling, i.e., including or not RC phase shifters. With the phase shifters deactivated, the measurement agrees very well with the simulation at offset frequencies beyond 100 kHz. Below this offset frequency, the slightly increased $1/f$-noise corner can be observed again. The measurements confirm that activating the RC phase shifters reduces the phase noise by about 2 dB at 1 MHz offset. At lower offset frequencies, the phase noise reduction even reaches 4 dB due to the suppression of flicker noise up-conversion [85]. However, the reduction of flicker noise up-conversion is approximately 3 dB lower than estimated by the post-layout simulations. This indicates that the RC phase shift deviates from the targeted 45°, possibly caused by a too pessimistic extraction of the parasitic capacitance.

It is worth observing that both architectures achieve a sufficient phase noise performance for WBAN applications such as BLE, which requires $-92\,$dBc/Hz at $1.5\,$MHz offset. Interestingly enough, although the quality factor of the QVCO inductors (11.5) is smaller than that of the VCODIV2 inductor (14.4), the QVCO achieves better phase noise performance at large frequency offsets than the VCODIV2. This seeming inconsistency is, however, in good agreement with theoretical analysis that predicts a 3 dB superior phase noise performance of a QVCO with respect to a single-tank VCO with equal tank properties [125, 130].

Concerning power consumption, the QVCO is much more efficient than the VCODIV2. While the former only requires $210\,\mu$A of supply current, the latter needs $335\,\mu$A, 45 % of which are contributed by the VCO and the remaining 55 % by the divider-by-2 circuit. On the other hand, the VCODIV2 architecture occupies a much smaller die area than the QVCO topology, namely $0.062\,$mm^2 compared to $0.169\,$mm^2, which is obviously related to the number and size of the inductors employed by the respective implementations.

Table 4.1 summarizes the performance of both quadrature generators. Due to the lack of a single-sideband mixer in the experimental setup, the quadrature error could not be properly measured. Oscilloscope measurements at the I- and Q-output confirm quadrature operation but with an increased phase error, which is mainly caused by the inadequate test setup. Nevertheless, postlayout simulations including mismatch effects demonstrate that the QVCO with phase shifters obtains similar phase errors, in the order of $-2.5°$, as the VCODIV2. It is also observed that the phase error performance of the QVCO slightly worsens by approximately 1° when the RC phase shifters are enabled because the magnitude of the coupling signal is reduced by the attenuation of the passive phase shifters.

Table 4.1 Measured performance of the presented quadrature generators (test chips 1a and 1b)

Parameter	VCODIV2	QVCO[a]
Supply voltage (V)	1.0	1.0
Oscillator current (μA)	150	210
Divider current (μA)	185	N/A
Digital tuning range (MHz)	270	190
Analog tuning range (MHz)	115	95
Output phase noise	-83.8@100 kHz	$-87.5\,/\,-84.2$ @100 kHz
(dBc/Hz)	-110.2@1 MHz	$-111.9\,/\,-110.2$ @1 MHz
	-118.4@2.5 MHz	$-120.3\,/\,-118.6$ @2.5 MHz
IQ imbalance		
Gain error (dB)	0.1[b]	0.3 / 0.4[b]
Phase error (°)	-2.5[b]	$-2.3\,/\,-1.1$[b]
Active area (mm^2)	0.062	0.169

[a]Performance with/without phase shifting

[b]Postlayout simulation at the worst-case corner including mismatch effects (no magnetic coupling considered between the two QVCO inductors)

Table 4.2 Performance comparison to recent low power quadrature generators

Parameter	[95]	[72]	[103]	[99]	This work	
CMOS process (nm)	180	180	130	180	90	90
Architecture	VCODIV2	VCODIV2	QVCO	QVCO	VCODIV2	QVCO[a]
Supply voltage (V)	1.8	0.6	1.0	1.2	1.0	1.0
Frequency (GHz)	2.2	2.2	2.2	2.4	2.4	2.4
Power (mW)	0.9	5.17	0.6	1.2	0.335	0.21
Phase noise at 1 MHz (dBc/Hz)	−117.0	−119.0	−110.7	−103.7	−110.2	−111.9
FoM_{LO} (dB)	184.3	178.6	179.8	170.5	182.9	186.1
Active area (mm^2)	0.18	≈0.45	≈0.4	≈1.0	0.06	0.17

[a] With RC phase shifters enabled

However, in case of the QVCO, the elevated measured phase error may also be due to another parasitic effect that has not been considered so far, namely magnetic coupling between the two inductors. This aspect is considered in the Sect. 4.1.2.2, which also demonstrates a possible solution in a further silicon implementation.

Table 4.2 compares the two designs to recently published work. The oscillator figure-of-merit (FoM$_{LO}$), defined as [11]

$$\mathrm{FoM_{LO}}(\Delta\omega) = 10\log\left[\left(\frac{\omega_0}{\Delta\omega}\right)^2 \frac{1}{\mathscr{L}(\Delta\omega)}\frac{1\,\mathrm{mW}}{P}\right], \qquad (4.3)$$

relates the phase noise $\mathscr{L}$ at an offset frequency $\Delta\omega$ to power dissipation P and oscillating frequency ω_0. It is worth observing both generators outperform previous works in power consumption and in case of the QVCO also in terms of FoM$_{LO}$. Figure 4.10 depicts the measured FoM$_{LO}$ at 1 MHz offset of the implemented quadrature generators versus power consumption. It has been measured by sweeping the bias current of the oscillator cores. For low bias currents, the phase noise performance

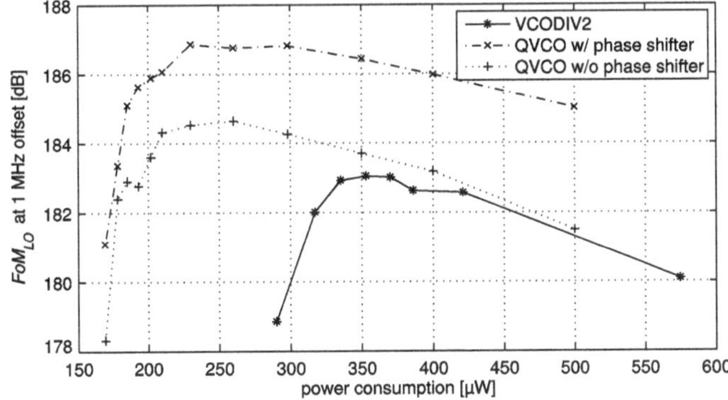

Fig. 4.10 Measured FoM of the presented quadrature generators

increases rapidly due to the increasing oscillation amplitude in the core. However, the oscillation swing is ultimately limited by the rails due to the complementary current-reuse architecture. Therefore, the phase noise performance does not improve anymore when such limiting effects come into play leading to a decay of the FoM_{LO} at higher power consumption.

4.1.1.4 Conclusion

To select the appropriate quadrature generation topology, two alternatives have been implemented in 90 nm CMOS, namely a VCO with divider-by-2 and a QVCO. Both topologies have been optimized for minimum power consumption. The QVCO obtains a significantly lower power consumption and, due to the passive RC phase shifters in the coupling path, it also achieves a slightly lower phase noise than the VCODIV2. On the other hand, the QVCO occupies a much larger silicon area due to the two required spiral inductors. Nevertheless, the QVCO is selected as the preferred topology for the transceiver since the LO power consumption is crucial in the passive receiver architecture.

4.1.2 QVCO with Direct Modulation

Having selected the appropriate oscillator topology, the previously designed QVCO will be used as a starting point for the design of the local oscillator of the transceiver. In this context, the QVCO requires two important changes with respect to the previously presented stand-alone QVCO (Fig. 4.4), namely an additional FSK modulator cell in the tank and bias switches to allow for single-core operation in transmit mode. The modified QVCO schematic is shown in Fig. 4.11. Another aspect taken into account for the new QVCO is the effect of magnetic coupling between the two core inductors, which is represented by the magnetic coupling factor k_{mgn} in Fig. 4.11. This effect may cause quadrature errors and hence needs to be compensated using a cancelation network, which is detailed in Sect. 4.1.2.2. Moreover, the transceiver is implemented in a 130 nm CMOS, while the previous designs have been realized in 90 nm CMOS. Note that this seeming step backward in minimum gate length has little effect regarding the active part of QVCO, because the 90 nm-CMOS implementation did not use minimum-sized transistors in order to maximize the effective transconductance ($g_m - g_{ds}$). More important is the change from two thick top-metals to only one thick top-metal which eventually leads to a reduced QVCO core inductance of now $L = 10.2\,\text{nH}$, realized by a differential 6-turn inductor with a loaded quality factor of $Q_L = 13$.

Since in TX-mode only one core is needed, switches allow to steer all the bias current to the I-core in this mode of operation. Doubling the current in this core increases the oscillation amplitude to almost rail-to-rail swing and so allows to relax the power demand of the PA drivers that connect the I-core to the PA output stage. Overall, this

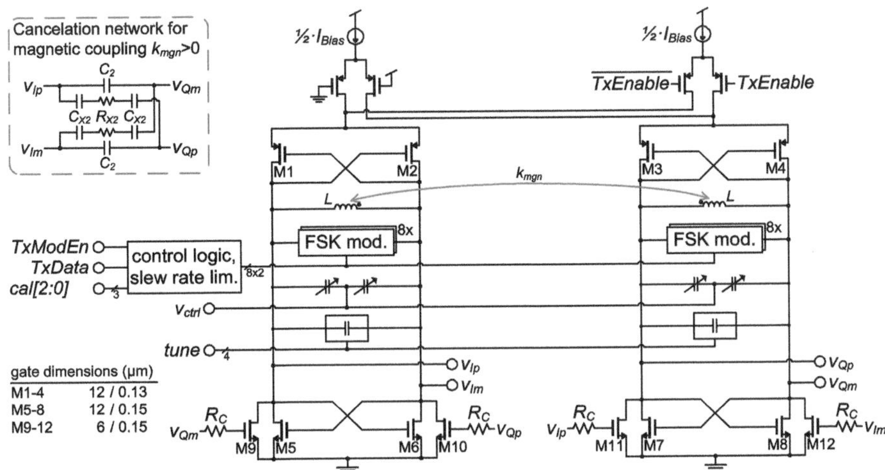

Fig. 4.11 Schematic of the final QVCO with FSK modulation cells and current steering switches in the bias path

results in a lower current consumption than if no current steering between the cores were applied. However, powering down the Q-core in TX-mode also requires a modification of the coupling transistor setup. While in the former QVCO both PMOS and NMOS transistors have been used for coupling devices, now the quadrature coupling is based on NMOS transistors only (M9-M12) but maintaining the same coupling coefficient of $m = 0.25$. To compensate the missing PMOS coupling transistors, the width of M9-M12 is increased accordingly. In TX-mode, the Q-core is not biased and the voltages $v_{Qp/m}$ decay to zero, and hence so do the gate voltages of M9/M10. Therefore, these NMOS transistors are off and do not affect the oscillation of the I-core when the Q-core is switched off.

4.1.2.1 Direct QVCO-Modulation

Direct QVCO modulation is realized by modulating the capacitance of the LC-tank [34, 52, 101]. The BLE data rate of 1 Mb/s and the FSK modulation index of nominally $h = 0.5$ impose a frequency deviation of $\Delta f = \pm 250\,\text{kHz}$ [3]. Since $\Delta f \ll f_c$, this frequency deviation implies a differential tank capacitance step ΔC given by,

$$\Delta C \approx \frac{2\Delta f}{f_c} C_{\text{tank}} = \frac{\Delta f}{2\pi^2 f_c^3 L} \tag{4.4}$$

where L and C_{tank} are, respectively, the inductance and total capacitance of the tank, and f_c represents the carrier frequency. It is worth observing the tiny values that ΔC can take. For instance, for the tank inductance of $L = 10.2\,\text{nH}$ used in this design, ΔC is only 85 aF, assuming $f_c = 2.44\,\text{GHz}$.

Such small capacitance changes for direct VCO modulation can be implemented by using an additional tank varactor. This, effectively, translates the required capacitance change into a voltage step that can be generated by means of a DAC [34]. The DAC approach also allows to shape the driving voltage in order to reduce the occupied bandwidth, but it incurs in increased power and area consumptions due to the required DAC and the following reconstruction filter.

Alternatively, the two control voltages that correspond to the two modulation frequencies can be stored on large memorization capacitors [52]. This solution allows for arbitrary and, in principle, accurate FSK modulation indices as they are set by the PLL. However, this approach leads not only to a large area overhead due to the capacitors and associated buffers but also to a longer start-up time because of the need for a calibration phase before modulation.

This design employs a digital approach inspired by the tuning strategy used in digitally controlled oscillators (DCO), proposed by Staszewski et al. [122]. The idea is not to adjust the analog control voltage of a varactor but to digitally select that incremental tank capacitor, among a finite set of instances, which better meets (4.4). The basic building block in this "spare box" approach is a simple PMOS transistor pair with the gates connected to the oscillating tank and the remaining terminals tied to a common control voltage v_{mod}. By switching this voltage from rail to rail, the PMOS transistor pair alternates between two states with a gate capacitance difference in the order of tens of aFs as requested in (4.4). This is illustrated in Fig. 4.12 for a small sized PMOS transistor pair in the selected technology. The proposed approach for direct QVCO modulation consists in exploiting the capacitance gap between these states to implement the requested frequency deviations.

In order to allow for a positive and a negative frequency deviation, two differential PMOS transistor pairs are required. This leads to the modulator cell shown in

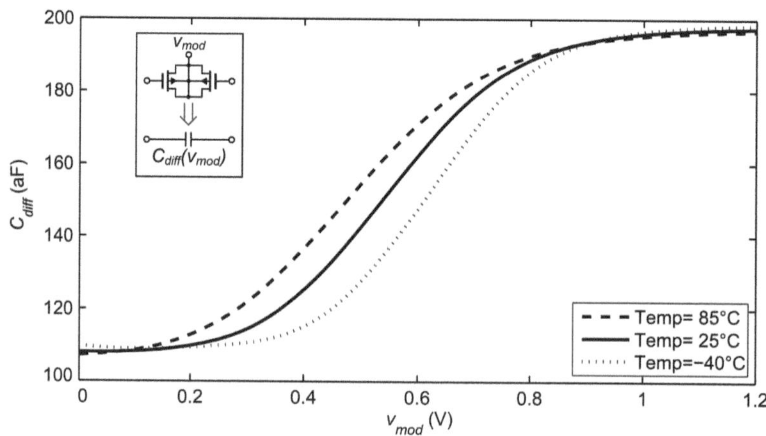

Fig. 4.12 Simulation of the differential capacitance C_{diff} of the PMOS pair shown in the inset ($w = 225$, $l = 150$ nm) versus control voltage v_{mod} for a common mode voltage at differential nodes of 0.3 V

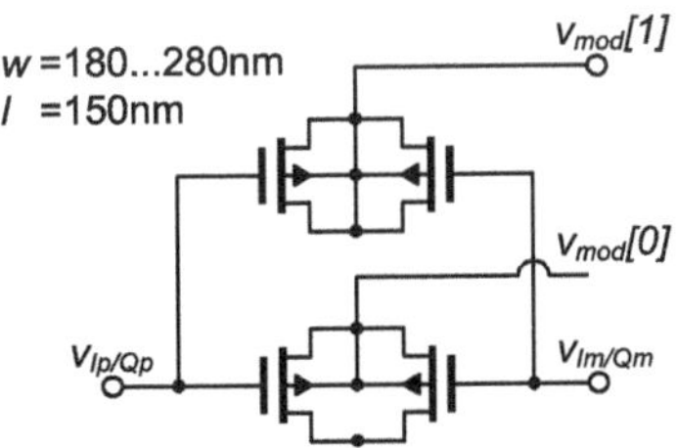

Fig. 4.13 Modulator cell for direct VCO modulation using standard PMOS transistors

Fig. 4.13. Setting both modulation inputs ($v_{mod}[1:0]$) to V_{DD} leads to a capacitance increment and hence to a negative frequency deviation. On the other hand, setting both inputs to ground results in a positive frequency shift. The intermediate state with one input to ground and the other one to V_{DD} is needed to set the QVCO to the carrier frequency when the PLL is closed.

To take care of the impact of process variations on the gate capacitance of PMOS transistors, five modulator cells with slightly different dimensions have been attached to each QVCO core. Hence, the gate width of the PMOS transistors ranges from 0.18 to 0.28 μm with 250 nm increments while the gate length is constant for all cells. By means of calibration, the modulator cell that better approximates the capacitance jump in (4.4) is selected for direct modulation, whereas the remaining cells have their v_{DSB} voltages tied to ground. Due to the small size of the PMOS transistors, the additional tank capacitance of this array is negligible with respect to the total tank capacitance. Note that the array of FSK modulator cells is functionally not needed in the Q-core, which is only active in RX-mode. However, it has to be physically implemented for symmetry considerations.

Self-calibration can be easily accomplished by using the divider of the PLL as a detector. The procedure is currently implemented off-chip by means of an FPGA which measures the time T_{cal} in which the divider output outruns the reference clock by one cycle. The actual modulation index is then calculated as $h = N \cdot T_{sym}/T_{cal}$, where N is the divider ratio, T_{sym} is the symbol period of 1 μs and the expected T_{cal} is around 250 μs. Note that this calibration has to be performed only once after start-up because it only addresses the static process-related uncertainties.

Concerning supply voltage variations, the digital switching characteristic of the PMOS $C-v$ curve is intrinsically robust due to the almost constant gate capacitance around $v_{mod} = 0.0$ V and $v_{mod} = 1.0$ V. With respect to temperature variations, the simulations in Fig. 4.12 also show that within the industrial temperature range of -40–$85\,°C$ the FSK frequency deviation stays within $\pm 2\%$. Therefore, in contrast to process variations, no further calibration versus temperature or supply voltage fluctuations is needed for the proposed FSK modulator cell.

Switching the same capacitance step of ΔC for all channels in the 2.4 GHz ISM band also adds a deterministic error. The actual modulation index will be different at the two outermost channels of the band due to the different total tank capacitance. According to (4.4) and with R being the data rate, the actual modulation index h can be calculated as

$$h \approx \frac{\pi^2 L}{R} \cdot \Delta C \cdot f_c^3 \tag{4.5}$$

showing a cubic proportionality with the carrier frequency f_c. However, considering the two extreme carrier frequencies of 2.402 and 2.48 GHz, the resulting error of the modulation index is only about 5 % and so well below the 10 % limit of the BLE standard. Moreover, as this error is deterministic it can also be taken into account by the calibration.

The modulating vector $v_{mod}[1:0]$ is generated by a simple digital control logic, whose outputs are slew-rate limited to remove the high-frequency content of the rectangular waveforms. The resulting trapezoidal switching, with rise- and fall times of approximately 400 ns, reduces the sidelobes of the output spectrum and so helps to comply with the BLE spurious emission requirements.

4.1.2.2 Magnetic Coupling Cancelation in the QVCO

The parasitic magnetic coupling between the two inductors of the QVCO is a source of quadrature error [138]. If the two inductors are located close to each other their magnetic fields interact with each other. To estimate the impact of this interaction on the quadrature phase error $\Delta\psi$, let us consider the physical layout of the two inductors of the previously designed QVCO as shown in Fig. 4.14a. To estimate the interaction, expressed as the magnetic coupling factor k_{mgn}, the two-inductor configuration has been simulated with the 3-dimensional electromagnetic solver Momentum, resulting in a k_{mgn} of 1.8 %. In order to work with positive k_{mgn}, we have to define the inductors on schematic-level with inverted polarities, as shown in Fig. 4.14b. Then, assuming small phase errors, the effect of magnetic coupling may be translated into a mismatch of quality factors of the two tanks

$$Q_{tank,I/Q} \cong Q_{tank} \cdot \frac{1}{1 \pm k_{mgn} \cdot Q_{tank}} \tag{4.6}$$

where Q_{tank} is the nominal tank quality factor. Using the theoretic analysis of Mirzaei et al. [85], which calculates the quadrature phase error as a function of tank mismatch,

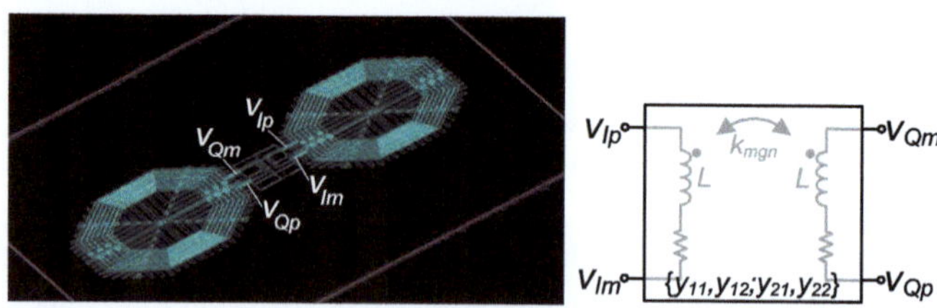

Fig. 4.14 QVCO inductor parameter extraction. **a** Physical layout. **b** Resulting polarity for the simplified model

the phase error can be estimated as

$$\Delta \psi = -k_p \cdot Q_{\text{tank}} \cdot \frac{(1 + m_0 \cdot \sin \phi_0) \cdot \cos \phi_0}{m_0 + \sin \phi_0} \tag{4.7}$$

where ϕ_0 is the ideal phase shift in the coupling path and m_0 is the transconductance ratio between the cross-coupled core transistors and the parallel coupling transistors. Therefore, with the polarity defined before, a magnetic coupling factor $k_{\text{mgn}} > 0$ pulls the cores toward in-phase operation.

Finally, considering that in the proposed QVCO the phase shift in the coupling path is realized by an RC-network, we have to take into account the corresponding attenuation with the following substitutions

$$\phi_0 \Rightarrow \phi_{RC} \tag{4.8}$$

$$m_0 \Rightarrow m \cdot 1/\sqrt{1 + \tan^2 \phi_{RC}} \tag{4.9}$$

where ϕ_{RC} is now the phase shift caused by the RC phase shifters in the coupling path ($\tan \phi_{RC} = 2\pi f_0 R_{\text{cpl}} C_{G,\text{cpl}}$) and m is still the transconductance ratio between core and coupling transistors. Combining (4.7–4.9) then yields the phase error $\Delta \psi$ as a function of the magnetic coupling factor k_{mgn} for the QVCO with RC-phase shifters

$$\Delta \psi = -k_{\text{mgn}} \cdot Q_{\text{tank}} \cdot \frac{1 + m \cdot \sin \phi_{RC} \cdot \cos \phi_{RC}}{m + \tan \phi_{RC}}. \tag{4.10}$$

In this QVCO design, the parameters ϕ_{RC} and m are nominally $\phi = 45°, m = 0.25$. Then, with $Q_{\text{tank}} = 13$ and $k_{\text{mgn}} = 1.8\,\%$, the estimated phase error is as high as $\Delta \psi = -12°$. In order to not degrade the zero-IF receiver by more than $1\,\text{dB}$, this phase error must be reduced to $|\Delta \psi| < 5°$.

One possibility to reduce this error is to simply increase the physical distance. However, to obtain a significant improvement the distance should be much larger than the diameter of the inductors [138]. This also increases the parasitic capacitance on the interconnects and hence the power consumption. Another possible solution is to use four inductors instead of two which allows for full symmetry with respect to the four quadrature phases [13]. In this case, single-ended inductors are used instead of differential ones, which leads to smaller inductances and hence also to an increased power consumption. The advantage of both these solutions is that they do not require the exact knowledge k_{mgn}, which is often difficult to predict.

However, in many cases k_{mgn} can be estimated from electromagnetic simulations or even from a measurements of a previous versions making it also possible to cancel the magnetic coupling. To this end, let us consider the two inductors as a two-port network which is described in terms of its admittance matrix (y-matrix), as indicated in Fig. 4.14b. Here the interaction between the two inductors is represented by the transfer admittances y_{12} and y_{21}. Therefore, adding a shunt network with transfer admittances of equal magnitude and opposite signs cancels the coupling. Figure 4.15a

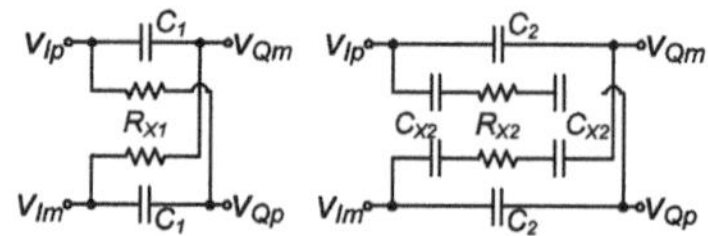

Fig. 4.15 Magnetic coupling cancelation networks. **a** Simplest network with four elements. **b** Improved cancelation network with eight elements

shows the simplest configuration of such a network with

$$C_1 = \frac{2\Im\mathfrak{m}\{y_{21}\}}{\omega_0} \approx \frac{2k_{\mathrm{mgn}}}{\omega_0^2 L} \tag{4.11}$$

$$R_{X1} = \frac{-1}{2\Re\mathfrak{e}\{y_{21}\}} \approx \frac{Q_L \omega_0 L}{4k_{\mathrm{mgn}}}. \tag{4.12}$$

Unfortunately, for the given design values ($L = 10\,\mathrm{nH}$, $Q_L = 13$, $k_{\mathrm{mgn}} = 1.8\,\%$, $\omega_0 = 2\pi \cdot 2.45\,\mathrm{GHz}$) the required value of R_{X1} is approximately $30\,\mathrm{k\Omega}$, which would add an excessive parasitic capacitance. In order to reduce the required resistance to values suitable for GHz-frequencies, the network of Fig. 4.15b can be used instead.

$$C_{X2} = \frac{-4}{\omega_0 \Im\mathfrak{m}\{1/y_{21}\}} \approx \frac{4k_{\mathrm{mgn}}}{\omega_0^2 L} \tag{4.13}$$

$$R_{X2} = -\frac{1}{2}\Re\mathfrak{e}\left\{\frac{1}{y_{21}}\right\} \approx \frac{\omega_0 L}{k_{\mathrm{mgn}} Q_L} \tag{4.14}$$

$$C_2 = \frac{2\Im\mathfrak{m}\{y_{21} - 1/(y_{21}^{-1})^*\}}{\omega_0} \approx \frac{4k_{\mathrm{mgn}}}{\omega_0^2 L} \tag{4.15}$$

Clearly, this cancelation technique is both frequency dependent and susceptible to process variations. Therefore, it may be used in narrow-band systems and only to reduce the phase error due to magnetic coupling to a tolerable level. Figure 4.16 shows that the decoupling network achieves at least a 11.5 dB reduction of the coupled voltage from one inductor to the other (both resonating at 2.45 GHz) for the worst-case process corners, which are mainly defined by the MOM capacitance variation. The statistical Monte Carlo simulation considering both process and mismatch variation shows that for most cases the magnetic coupling is reduced by more than 19 dB. This reduces the expected phase error due to magnetic coupling from $-12°$ to values within $\pm 3°$.

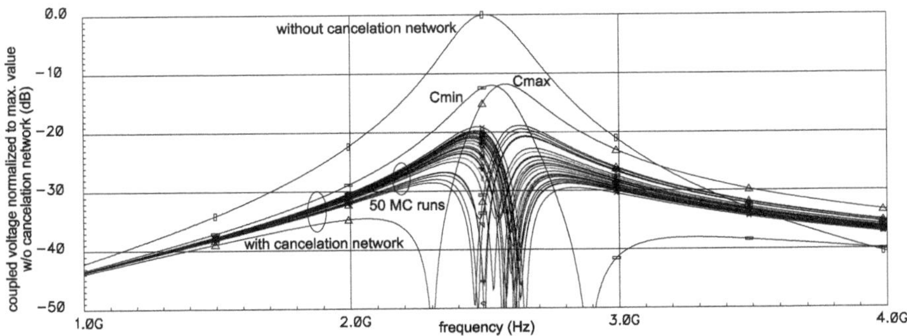

Fig. 4.16 Monte Carlo (MC) and worst-case simulation of magnetic coupling cancelation (normalized to uncanceled coupling)

4.1.3 Finite-Modulo Fractional-N PLL with Spur Compensation

The carrier frequency is controlled by means of a conventional type-II, third-order charge pump PLL [106], as shown in Fig. 4.17. To suppress near-by frequency pulling effects it implements a high loop bandwidth of $f_{bw} = 1\,\text{MHz}$ and hence requires also a high reference frequency of $f_{ref} = 20\,\text{MHz}$. The PLL employs a simple single-ended charge pump (CP) with tristate operation. It is formed by two complementary current sources providing a current of $80\,\mu\text{A}$. The output drains of these current sources are connected to the CP output node by means of two switches, which are controlled by the phase/frequency detector (PFD) of the PLL. Hence, no current

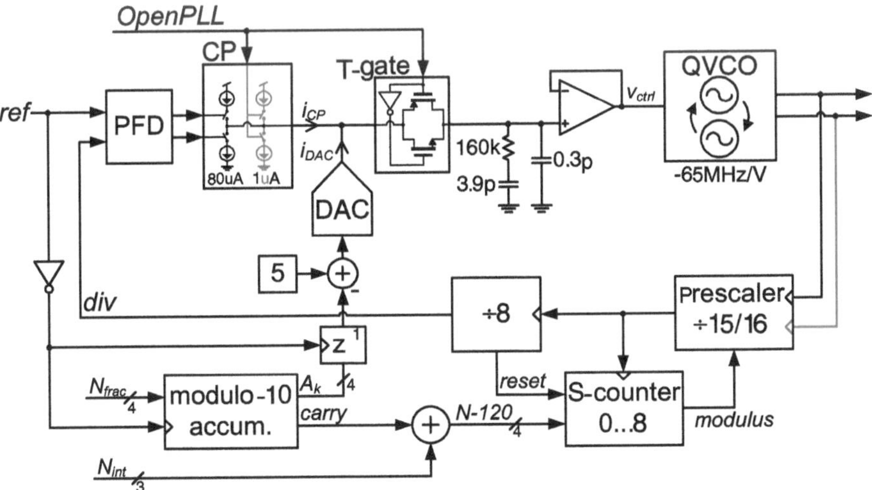

Fig. 4.17 Schematic of the implemented fractional-N PLL with DAC-based spur compensation

circulates through the output branch of the CP when both PFD outputs are low. This reduces the average power consumption of the CP output branch to less than $10\,\mu W$. The reference spurs at $20\,MHz$ offset are sufficiently attenuated by the low-pass characteristic of the loop.

During direct modulation, a complementary transmission gate (T-gate) opens the PLL and so disconnects the CP from the loop filter. With the PLL opened, the charge pump keeps its output voltage at approximately half the supply voltage using auxiliary $1\,\mu A$ current sources (depicted in gray in Fig. 4.17). Therefore, both transistors in the T-gate are reversely biased and hence operate deeply in their cut-off region. This reduces the leakage current from the loop filter capacitance to the T-gate and keeps the frequency drift during direct modulation well below the BLE limit of $400\,Hz/\mu s$ [145]. On the other side, the loop filter is connected to a PMOS input voltage buffer to prevent leakage toward the QVCO. This buffer adds a pole to the transfer function of the PLL at about $10 \cdot f_{\mathrm{bw}}$, thus reinforcing the reference spur filtering action.

The programmable divider employs a pulse-swallow architecture to provide integer divide ratios from 120 to 128. For fractional-N functionality, this divide ratio is modulated by the *carry* output of an accumulator that overflows at a value of 10, as shown in Fig. 4.17. Accordingly, the average divide ratio $\overline{N}$ amounts

$$\overline{N} = 120 + N_{\mathrm{int}} + N_{\mathrm{frac}}/10 \qquad (4.16)$$

and so provides the required $\frac{1}{10}$ resolution to synthesize the $2\,MHz$ BLE channel spacing from the reference frequency of $20\,MHz$.

The power consumption of high-speed frequency dividers is mainly defined by the sub-blocks that operate at the highest frequency, i.e., the input frequency. Therefore, a phase-switching prescaler is employed that contains only one asynchronous divider-by-2 operating at the high input frequency [40]. All subsequent dividers already operate at divided frequencies and hence consume less. Figure 4.18 shows the implemented 15/16-prescaler. The two cascaded divide-by-2 blocks at the input of the prescaler consist of DSTC latches in a master-slave flip-flop configuration

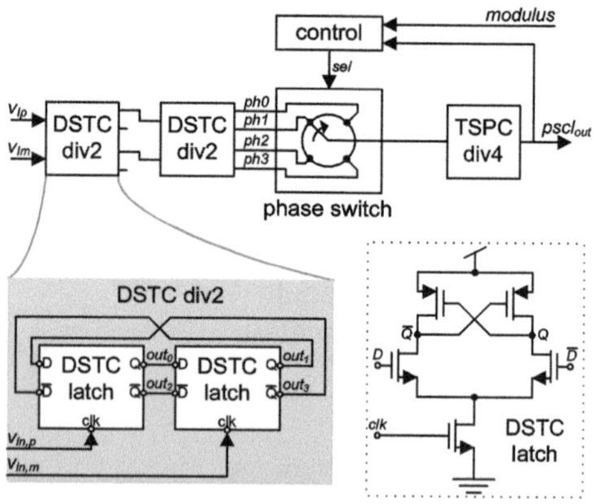

Fig. 4.18 Phase-switching prescaler (modulo 15/16)

[148]. The DSTC-based divider-by-2 achieves a low power consumption due to the few internal nodes and hence low parasitic capacitance. The phase switch is implemented using four tri-state buffers, where only one is activated at the same time. After the phase switch only single-ended signals are required, and hence the following divider-by-4 uses two cascaded true-single-phase-clocked (TSPC) flip-flops [128]. Upon rising output edges of the prescaler, the phase switch is rotated toward the preceding phase which leads to a division by 15 if the modulo input is set accordingly. Otherwise, the phase switch maintains its state leading to a division by 16.

Note that the phase switching prescaler is driven by the two differential inputs of the I-core only, but for symmetry reasons also the Q-core has to be loaded with a DSTC-based divider-by-2 (not shown in Fig. 4.18). Nevertheless, since this dummy divider is always powered down, the load conditions are not perfectly symmetrical for I- and Q-core. However, the resulting phase error due to this effect is less than $1°$ according to post-layout simulations, which does not require any countermeasure.

The modulation of the divider produces periodical patterns of erroneous, but deterministic, charge pump pulses which give rise to spurs at multiples of $f_{\mathrm{ref}}/10$ around the output frequency of the PLL. The width of these predictable pulses are integer multiples of $\Delta t = T_{\mathrm{LO}}/10$, where T_{LO} is the oscillation period of the QVCO. In order to cancel the generation of spurs arising from the modulation process, a simple time-domain compensation technique based on digital-to-analog conversion (DAC) is employed [126]. The operation principle is illustrated in Fig. 4.19 and essentially consists in the injection of current pulses which ideally cancel the charge erroneously delivered by the CP. The width of these current pulses is constant and equal to $T_{\mathrm{ref}} = 1/f_{\mathrm{ref}}$; however, their amplitudes, provided by a current steering 4-bit DAC, vary in terms of the delayed output of the modulo-10 accumulator, A_{k-1}, as

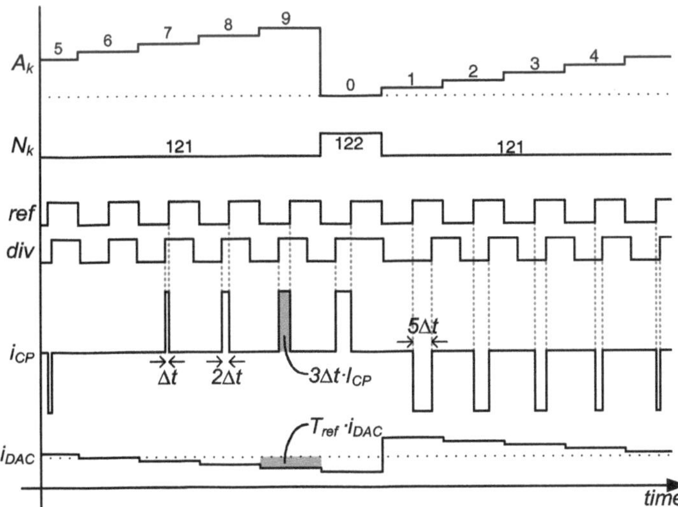

Fig. 4.19 Illustration of the fractional spur compensation for an average divide ratio $\overline{N} = 121.1$

$$i_{\mathrm{DAC}} = (5 - A_{k-1}) \cdot \frac{I_{\mathrm{CP}} \cdot T_{\mathrm{LO}}}{10 \cdot T_{\mathrm{ref}}} \approx (5 - A_{k-1}) \cdot \frac{I_{\mathrm{CP}}}{1220} \qquad (4.17)$$

where I_{CP} is the 80 µA charge pump current and the divide ratio N is approximated by 122, which corresponds to the channel at the center of the ISM band. The error due to this approximation is less than $\pm 2\%$ and allows for sufficient spur suppression, as it has been confirmed with time-domain simulations of the PLL.

4.1.4 Experimental Results

The proposed PLL with direct QVCO-modulation has been implemented together with a first version of the transceiver frontend (see Sect. 4.2) in a 130 nm CMOS process with six metal layers including one thick top metal layer (test chip 2, Fig. 4.20). Note that the transformer of the transceiver frontend and the inductors of the QVCO are placed well apart to reduce the magnetic coupling between them (less than 0.2 % according to our estimations) and hence also reduce frequency pulling effects. The measurements have been carried out with the chip assembled in a QFN36 package and soldered onto a test board and with a nominal supply voltage of 1 V.

4.1.4.1 RX-Mode

In RX-mode both QVCO cores are active with a bias current of 200 µA per core. According to post-layout simulations, this bias current yields a peak-to-peak oscillation voltage of approximately 250 mV on each of the four quadrature phases. The measured supply current of the PLL is 270 µA, where the dominant portion is consumed by the prescaler (≈ 200 µA).

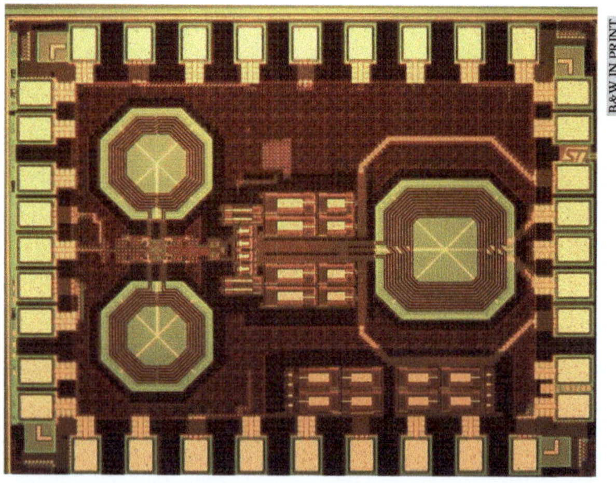

Fig. 4.20 Micro photograph of the frequency synthesizer with TRX frontend prototype (test chip 2) occupying a die area of 1.4 by 1.1 mm (1.54 mm^2). The synthesizer with QVCO and fractional-N PLL are at the *left* hand side and the transformer-based TRX frontend on the *right*

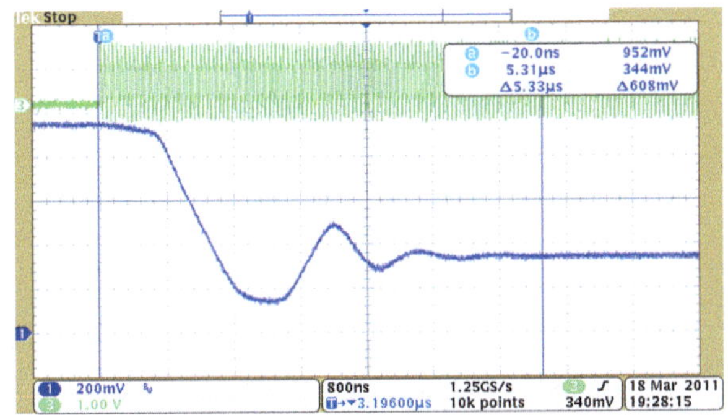

Fig. 4.21 Measured start-up behavior of the PLL: 20 MHz reference clock (*top*) and the QVCO control voltage v_{ctrl} (*bottom*)

The start-up timing of the fractional-N PLL is shown in Fig. 4.21. A fast settling within 5.5 µs is observed, due to the 1 MHz loop bandwidth of the PLL. This is well below the minimum interframe spacing of $T_{\text{IFS}} = 150$ µs defined by the BLE standard, which allows to switch off the synthesizer between two data packets in order to save power.

Figure 4.22 shows the measured power spectral density (PSD) of the locked PLL in RX-mode, i.e., with both QVCO cores activated, for a fractional division $\overline{N} = 121.1$. As can be seen, by using the DAC-based cancelation technique described in Sect. 4.1.3, the fractional spur at 2 MHz offset is attenuated by 18.6 dB, in the example. Indeed, this cancelation scheme is able to keep the fractional spurs at 2 and 4 MHz offset for all the BLE channels below −37 and −46 dBc, respectively.

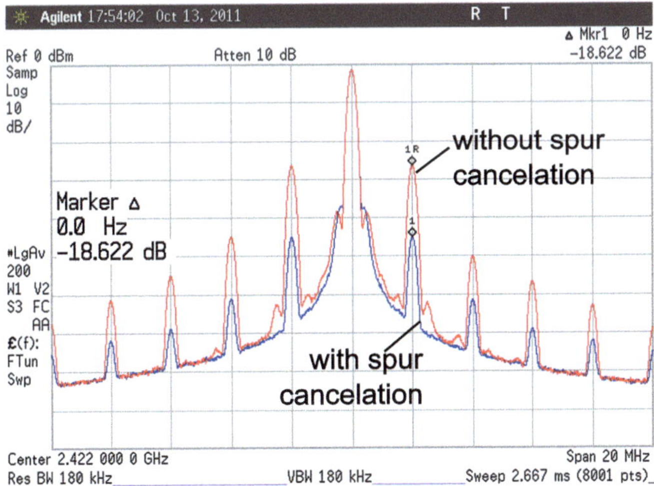

Fig. 4.22 Measured PLL output spectrum with both QVCO cores activated with and without spur cancelation

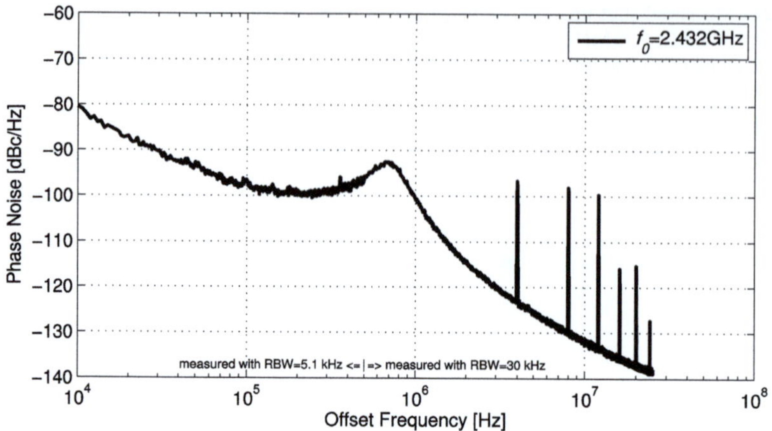

Fig. 4.23 Measured phase noise of the PLL in RX-mode

Figure 4.23 shows the phase noise of the synthesizer. In RX-mode, both QVCO cores are active and the fractional-N PLL is closed. Hence, the close-in phase noise is reduced by the PLL at offset frequencies below 500 kHz. Contrarily, around the loop bandwidth of $f_{bw} = 1\,\text{MHz}$, the phase noise level peaks due to the noise contribution of the charge pump and the loop filter, while at larger offset frequencies the QVCO phase noise is dominant. Moreover, the measured LO spectrum, obtained for $\overline{N} = 121.6$, exhibits a reference spur at 20 MHz as well as fractional spurs at integer multiples of 4 MHz due to the even fractional value of $N_{frac} = 6$.

Next, the phase accuracy of the LO is measured using the down-conversion mixer of the provisional TRX frontend. Figure 4.24 shows that the quadrature amplitude error keeps within $\pm 0.4\,\text{dB}$ and the phase error remains below $1.5°$ along the complete

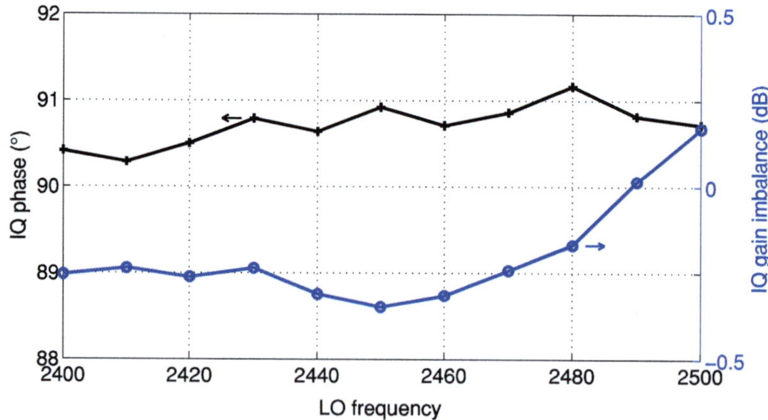

Fig. 4.24 IQ imbalance of the LO with magnetic coupling cancelation (chip 3), measured at down-converted baseband outputs of the TIA

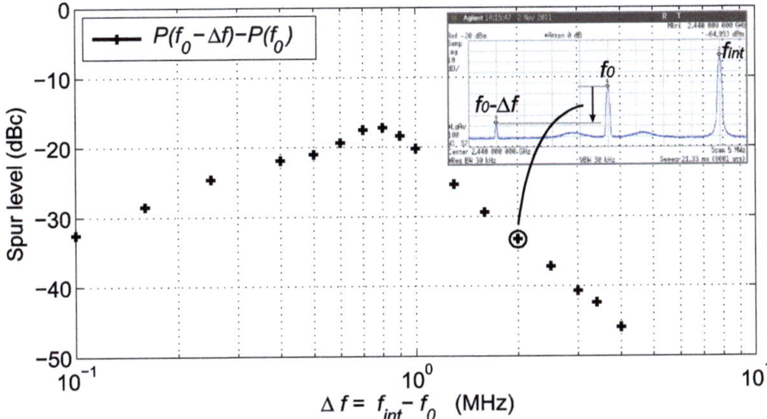

Fig. 4.25 Measured LO-pulling with an interferer at $f_{int} = f_0 + \Delta f$ with $P_{in} = -10\,\text{dBm}$, which causes the PLL to generate an LO spur at $f_0 - \Delta f$

ISM band. This is significantly lower than the $12°$ quadrature phase error theoretically calculated with (4.10) for a magnetic coupling of $k_{mgn} = 1.8\,\%$ and confirms the efficiency of the cancelation network for reducing the IQ imbalance of the LO described in Sect. 4.1.2.2.

Since the incoming signals and the LO of the transceiver operate in the same frequency range, the QVCO is potentially susceptible to frequency pulling effects. To evaluate this effect, the LO has been measured under the influence of a strong interferer at a near-by frequency f_{int}. Such interferer pulls the LO frequency toward f_{int} while the PLL forces the LO to output an average frequency f_0 as programmed by the divide ratio. These conflicting trends create a spur in the sideband of the LO opposite to f_{int}, which can be used as a measure of the LO pulling [107]. Figure 4.25 plots the relative power of these spurs versus the offset frequency $\Delta f = f_{int} - f_0$ for an interferer input power of $-10\,\text{dBm}$ at the antenna input of the transceiver frontend, which will be described in detail in the following Sect. 4.2. To avoid saturation of the RX-path at this power level, which is also the maximum tolerated by BLE, the input attenuator of the frontend is set to maximum attenuation for this measurement. It can be observed that the LO pulling peaks approximately at the loop-bandwidth of the PLL of $1\,\text{MHz}$. For small offset frequencies, the PLL suppresses the pulling effect whereas, for large offsets, the QVCO becomes less susceptible to the interferer due to the increased difference between the tank resonance and the interferer frequency. The spur levels of -21, -34, and $-42\,\text{dBc}$ at interferer offset frequencies of 1, 2, and $3\,\text{MHz}$, respectively, are low enough to comply with the BLE requirements on interference blocking.

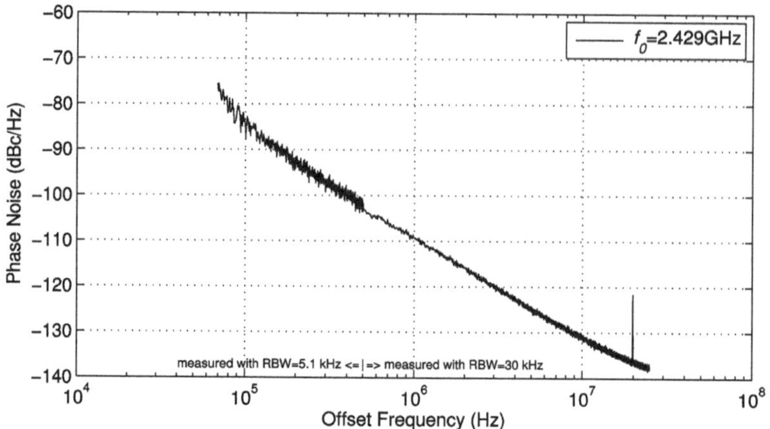

Fig. 4.26 Measured free-running single-sideband phase noise of the QVCO with one active core using two different resolutions bandwidths (RBWs)

4.1.4.2 TX-Mode

In TX-mode, all the QVCO bias current is steered to the I-core and hence the measured supply current drawn by the QVCO is the same as in RX-mode, i.e., $400\,\mu A$. According to post-layout simulations, this bias current yields a peak-to-peak oscillation voltage of approximately $450\,mV$ on the two outputs. The divider is powered down during data transmission since the PLL is opened during that period of time. However, for test purposes, the divider may be activated also in TX-mode in this implementation in order to monitor the output frequency by using the divider as a sensor, as will be explained later in this sub-section.

During data transmission, the PLL is not locked and, hence, only the free-running QVCO phase noise affects the output spectrum. Figure 4.26 shows the unlocked phase noise level of the single-core QVCO of $-118.5\,dBc/Hz$ at an offset of $2.5\,MHz$. Note that with the opened PLL, a reference spur at $20\,MHz$ is present which is caused by parasitic substrate coupling. However, its power level of $-71.2\,dBc$ is sufficiently low to easily comply with the BLE spurious emission requirements.

The control signals of the transmitter for opening the PLL and starting the modulation, as well as the actual data stream to be transmitted, have been synthesized using an FPGA to allow for a flexible test setup. The FSK modulation can be evaluated in time-domain by observing the delay between edges of the PLL reference clock, ref, and the output of the fractional-N divider, div. With the PLL closed, these two signals are phase-locked, whereas during open-loop modulation the divider acts as a sensor for the QVCO output frequency. Considering a QVCO output frequency of $f_0 = f_c + \Delta f(t)$ with the carrier frequency being set to $f_c = N \cdot f_{ref}$ and $\Delta f(t)$ being the FSK modulation portion, the delay t_{del} between the reference clock and the divider output clock can be expressed as

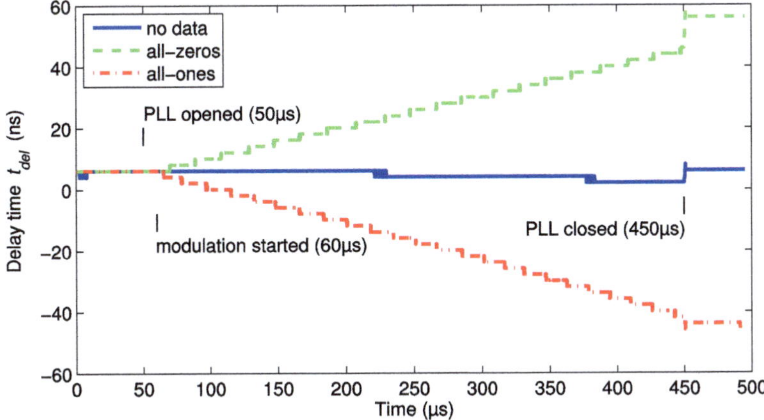

Fig. 4.27 Measured evolution of the delay time t_{del} between rising edges of the reference clock *ref* and divider output *div* when the PLL is opened ($f_c = 2.44\,\mathrm{GHz}$, cal $= 1$)

$$t_{\mathrm{del}}(t) = -\int \frac{\Delta f(t)}{f_c}\,\mathrm{d}t + t_0 \qquad (4.18)$$

where t_0 denotes the initial delay at the beginning of the integration. Figure 4.27 shows three different cases of the evolution of this delay when the PLL is opened for $400\,\mu\mathrm{s}$ and then closed again for $100\,\mu\mathrm{s}$. In the first case no data is sent, meaning that the FSK modulator remains in its intermediate state and $\Delta f(t) = 0$. Consequently, the delay stays almost constant throughout the opening time of the PLL. The other two cases show the linear evolution of the delay for an all-zeros ($\Delta f(t) = -250\,\mathrm{kHz}$) and an all-ones ($\Delta f(t) = +250\,\mathrm{kHz}$) data packet, as expected by (4.18). This property of integrating the actual frequency deviation into a delay time can also be utilized in a self calibrating scheme, which is currently not included on chip. Once the PLL is closed, the delay returns to an integer multiple of a reference clock period of 50 ns, with an offset t_0 caused by the delay of the I/O buffers.

Figure 4.28 shows the modulation index for the five calibration words, measured in frequency-domain using the Agilent E4440A PSA spectrum analyzer. It can be seen that for the calibration word cal $= 1$, the modulation index stays within the $0.5 \pm 10\,\%$ limit along the 2.4–2.48 GHz ISM band. For all calibration words, the typical frequency dependency as predicted by (4.5) can be observed.

The accuracy of the modulation index with respect to temperature variations is shown in Fig. 4.29. It reveals a positive temperature coefficient of approximately $6\,\%/100\,^{\circ}\mathrm{C}$, from -15 to $85\,^{\circ}\mathrm{C}$, which allows to keep the modulation well within the BLE limits. Regarding the impact of supply variations, it has been observed that the modulation index remains within $\pm 1\,\%$ from 0.9 to 1.25 V.

The effect of leakage is illustrated in Fig. 4.30, which shows the frequency drift rate versus temperature, which has been measured with the time-domain method described before. Beyond $50\,^{\circ}\mathrm{C}$, an exponentially rising drift rate is observed. This

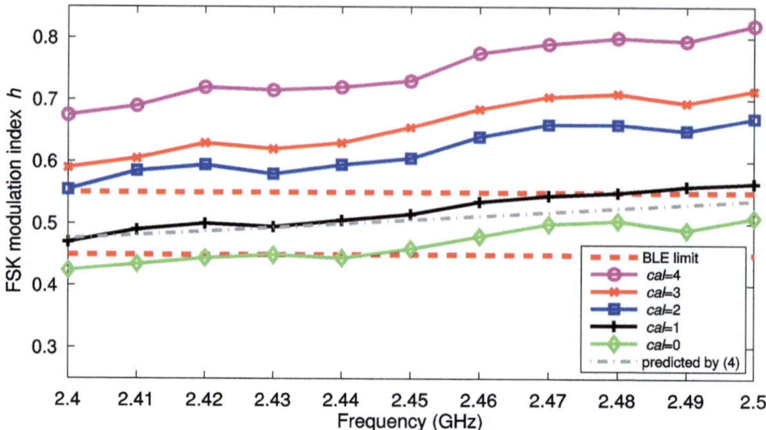

Fig. 4.28 Measured FSK modulation index h for all calibration values ($V_{DD} = 1.0$ V, $T = 25\,^{\circ}$C) and the calculated trend-line calculated with (4.5)

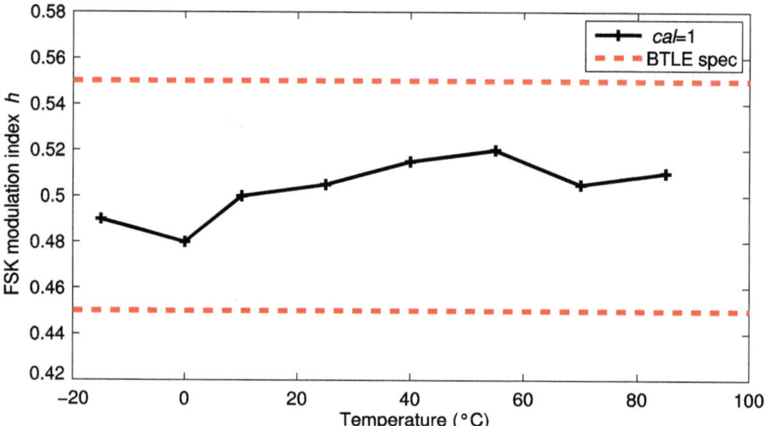

Fig. 4.29 Measured FSK modulation index h versus temperature for $cal = 1$ ($f_c = 2.44$ GHz, $V_{DD} = 1.0$ V)

is caused by the leakage current that discharges the loop filter capacitance. However, the drift rate stays well below 400 Hz/μs and meets the BLE requirements.

Figure 4.31 shows the PSD of the directly modulated LO in TX-mode, measured at the antenna port of the transceiver at the maximum output power of 1.6 dBm. Random data packets of 376 bits, the maximum length specified by BLE, are sent at a data rate of 1 Mb/s. Packet transmissions are interspaced by 10 μs, interval in which the PLL is closed and the carrier frequency is restored. The figure illustrates the effect of enabling/disabling the slew-rate limiter on the modulating signals $v_{\mathrm{mod}}[1:0]$. Note that, by shaping the modulation signals, the sidelobes are notably reduced and the margin to the BLE spurious emissions mask (dotted line in Fig. 4.31) increases by more than 10 dB for carrier frequency offsets >2 MHz.

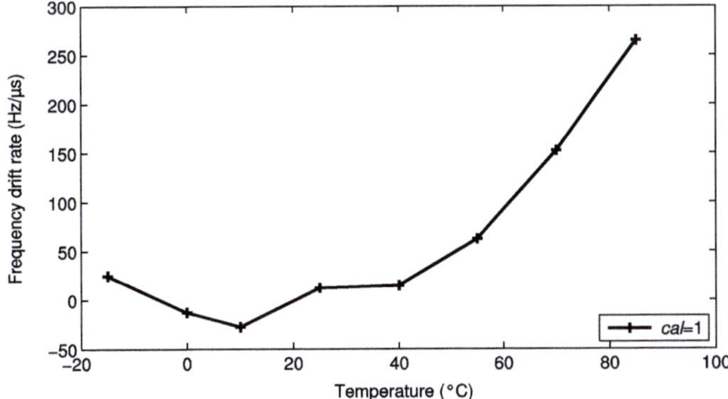

Fig. 4.30 Frequency drift rate f_{drift}/t versus temperature ($f_c = 2.44\,\mathrm{GHz}$, $V_{DD} = 1.0\,\mathrm{V}$)

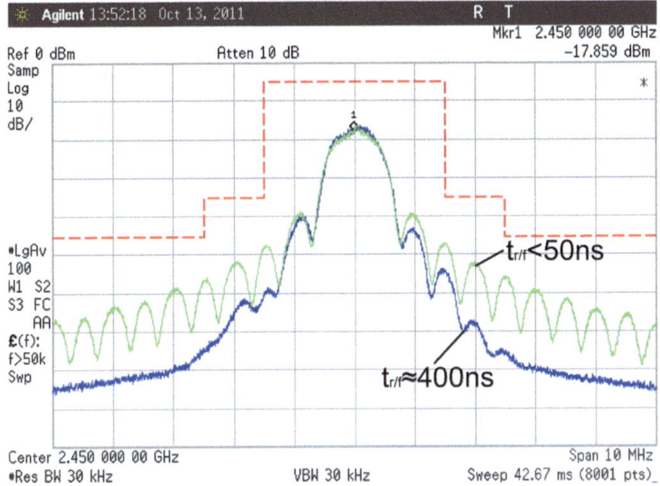

Fig. 4.31 Output spectrum of the frequency synthesizer at the maximum power level of the transmitter (1.6 dBm) with the slew rate limiter enabled ($t_{r/f} \approx 400\,\mathrm{ns}$) and disabled ($t_{r/f} < 50\,\mathrm{ns}$)

4.1.5 Conclusion

A low power frequency synthesizer has been presented in this section, which is tailored to the Bluetooth low energy standard. It is based on a QVCO that can be operated in two configurations according to the operation modes of the transceiver, i.e., in RX- and TX-mode. The performance of the frequency synthesizer is summarized in Table 4.3.

In RX-mode, both QVCO cores are active and hence generate the required quadrature phases. In order to reduce the magnetic interaction of the two QVCO inductors, a passive cancelation network has been proposed. With this cancelation network,

Table 4.3 Measured performance summary of the direct-modulation frequency synthesizer

Parameter	RX-mode	TX-mode
Technology	130 nm-CMOS	
Supply voltage (V)	1.0	
Current consumption (μA)	670	450
QVCO	400	
PLL	270	50
Frequency range (MHz)	$2400 + 2k(k = 1, 2, \ldots 41)$	
Settling time	$5.5\,\mu s$	
Phase noise	-98.3 @ 100 kHz	-88.2 @ 100 kHz
(dBc/Hz)	-100.3 @ 1 MHz	-110.9 @ 1 MHz
	-118.1 @ 2.5 MHz	-119.7 @ 2.5 MHz
Spur level	< -37 @ 2 MHz	N/A
(dBc/Hz)	< -46 @ 4 MHz	N/A
IQ imbalance		
Gain error (dB)	< 0.4	N/A
Phase error (°)	<1.5	N/A
GFSK modulation index	$0.5 \pm 10\% \, (-15\text{--}85\,°\text{C})$	
Frequency drift rate (Hz/μs)	$<300 \, (-15\text{--}85\,°\text{C})$	
Active area (mm^2)	0.35	

the measured phase error is less than $1.5°$. The center frequency of the synthesizer is set using a fractional-N PLL, which settles rapidly within $5.5\,\mu s$ and is able to synthesize carrier frequencies at a channel spacing of 2 MHz. Fractional spurs are attenuated using a simple DAC-based compensation scheme, which is sufficient to comply with the requirements of BLE.

In TX-mode, all the bias current is steered to one QVCO core as no quadrature signals are required. Again, the center frequency is set by the fractional-N PLL. However, in TX-mode the PLL is opened to directly modulate the QVCO tank capacitance. To this end, a modulator cell that employs the gate capacitance of a PMOS transistor is introduced to switch the QVCO tank capacitance. Because of its digital C–v characteristic, the PMOS-based modulator is intrinsically robust against supply voltage variations. The prototype has been successfully tested from -15 to $85\,°\text{C}$ and a supply voltage $1.0\,\text{V} \pm 10\%$ showing that the FSK frequency deviation stays well within the BLE limits of $250\,\text{kHz} \pm 10\%$. Also the frequency drift with the PLL opened complies with the BLE specification, which imposes a drift rate of less than $400\,\text{Hz}/\mu s$.

4.2 Low Power Transceiver Frontend

Before going into the details of the proposed low-power transceiver frontend, let us review once again the main requirements of the frontend. Concerning the transmit path, GFSK-modulation is already performed within the frequency synthesizer.

Hence, in the frontend, the modulated signal only has to be amplified to output approximately 1 mW to the antenna. To this end, an efficient power amplifier (PA) is needed because its power consumption will dominate the overall DC power draw in TX-mode. In the receive path, the frontend has to implement the frequency translation from radio to baseband frequencies by means of a quadrature down-conversion mixer. In the selected architecture, a passive RF frontend is employed, as proposed by Cook et al. [38], in order to minimize the power consumption in RX-mode. This is possible because the targeted sensitivity of -80 dBm allows for noise figure (NF) as high as 19 dB.

Concerning the TX, switching type class-E PAs are known for their good efficiency but also for their high nonlinearity. Therefore, as linearity is no issue for constant-envelope signals, class-E PAs are frequently used for the amplification of frequency- or phase-modulated signals [57, 69, 94], such as the GFSK-modulation specified by the BLE standard. However, while efficiencies as high as 65 % have been achieved for class-E amplifiers with an output power on the order of 1 W [31, 69], much lower values of around 10 % are usually reported for PAs with mW output power levels [26, 37, 57]. Generally, to design an efficient PA, the load impedance and supply voltage has to be chosen adequately for the desired output power. Assuming ideal conditions, the maximum output power $P_{\text{out,ideal}}$ of a differential class-E amplifier can be calculated from [111]

$$P_{\text{out,ideal}} = \frac{32}{\pi^2 + 4} \frac{V_{DD}^2}{R_L} \tag{4.19}$$

where V_{DD} and R_L are the supply voltage and differential load resistance, respectively. A fully differential structure has the advantage that it presents not only symmetric loading to the LO but also shifts the supply lines switching noise generated by the PA to twice the carrier frequency, where it can be filtered more efficiently [111]. According to (4.19), optimization of a class-E PA for low output power levels requires to either lower the supply voltage [33, 38], or increase the load resistance. In order to use a common supply for the complete transceiver, the former option is dismissed in this work and the supply voltage is set to 1.0 V. Using this constraint, and considering an output power of 0 dBm, the required differential load resistance is $R_L = 2.3\,\text{k}\Omega$. If losses in a practical implementation are taken into account, the required differential load resistance seen by the differential class-E stage decreases to values on the order of 1 kΩ.

The RX-path implementation also benefits from high RF impedance levels. First of all, up-conversion of the impedance also increases the received voltage, i.e., it provides passive voltage gain. Secondly, since the passive mixer is based on steering the RF signal by means of switches, we can tolerate a higher on-resistance if the RF impedance is higher than the standard 50 Ω. This leads to smaller switches, which reduce the loading of the QVCO that drive the switches.

4.2.1 Transformer

In the proposed transceiver frontend, a transformer is used to convert a differential antenna impedance of $100\,\Omega$ [2] to about $1\,\mathrm{k\Omega}$, as required for the PA load impedance. Although such an impedance conversion is also possible using an LC-matching network [38], the transformer solution is preferred here because it provides a wider bandwidth and also allows to connect a single-ended antenna by simply tying one of the antenna terminals of the transformer to ground [57]. To design a transformer with a high impedance on the internal coil, a large number of turns is required, similar to the QVCO described in the previous section. In both cases, the parallel loss conductance of the coil is minimized by employing the largest number of turns possible. Again, the number of turns is ultimately limited by the capacitive losses connected to the coil. Therefore, the transformer for impedance transformation also employs six turns at the internal coil. Figure 4.32 shows the implemented square symmetric step-up transformer with a turn ratio of 2:6. This turn ratio leads ideally to an impedance up-conversion by a factor of $3^2 = 9$, and hence to an internal RF impedance slightly below $1\,\mathrm{k\Omega}$. In order to reduce ohmic losses in the antenna coil, it consists of two parallel windings with two turns each. This also helps to improve the coupling factor of the transformer k_{tf} and, hence, reduces the insertion loss of the transformer [75].

The electrical behavior of the transformer has been characterized using a 3D electromagnetic solver. The resulting S-parameter set was then fitted into a direct form model of the transformer [75]. The most significant parameters of the model

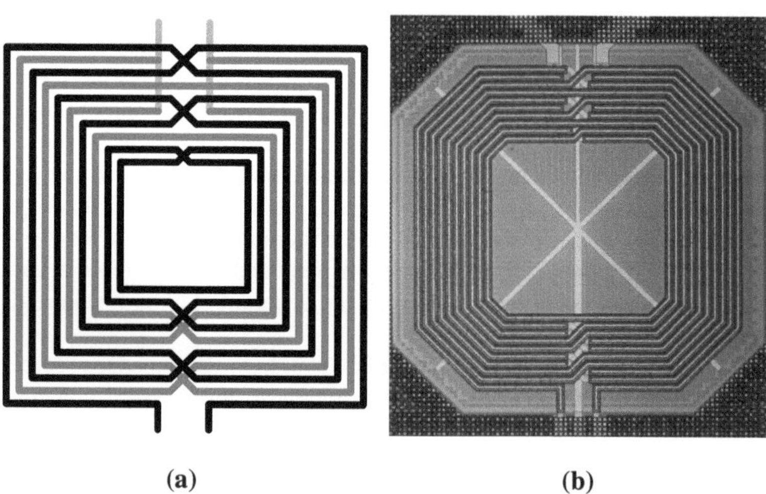

(a)   (b)

Fig. 4.32 Implemented 2–6 step-up transformer used for up-conversion of antenna impedance. **a** Illustration of the transformer, where the *gray* coil (two parallel windings with two turns each) connects to the antenna and the *black* coil (six turns) to the internal PA. **b** microphotograph of the transformer occupying an area of 340 by $340\,\mu\mathrm{m}$ (incl. shielding)

Table 4.4 Transformer parameters

Parameter	PA load coil	Antenna coil
Number of turns	6	2
Windings in parallel	1	2
Serial inductance (nH)	11.6	1.56
Serial resistance (Ω)	18.0	3.2
Coupling factor k_{tf}	0.85	
Sim. insertion loss @2.4 GHz	1.8 dB	
Area	0.116 mm^2 (340 × 340 µm)	

are presented in Table 4.4. According to this model, the transformer exhibits an insertion loss of 1.8 dB and converts the 100 Ω antenna impedance into an internal RF impedance of 840 Ω.

4.2.2 Frontend Implementation

Figure 4.33 shows the schematic of the proposed low-power RF frontend with the transformer at the antenna interface. In the transmitter, a switching-type, differential class-E PA operating in saturation is used for the constant envelope GFSK modulation specified in BLE [110]. The PA is implemented by a switching NMOS pair, M1-2 (Fig. 4.33). Two cascaded inverters act as driving stages and allow the gate voltages of M1-2 to swing from rail-to-rail. In order to provide output power programmability, the PA-drivers and the switching pair are cast into four equally-sized parallel branches. They are binary controlled so that 1, 2, or 4 branches can be jointly

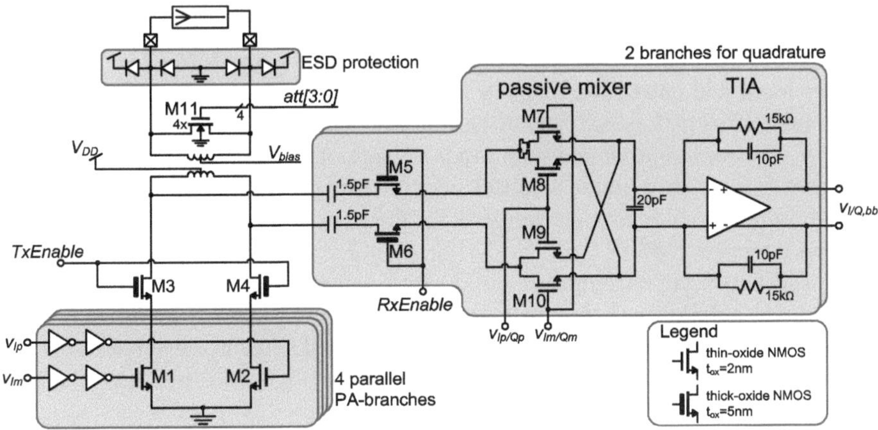

Fig. 4.33 Schematic of the transceiver frontend

activated. The cascade stage M3-4, common to the four branches, increases the output impedance of the PA when signal *TxEnable* is ON, and disconnects the PA from the transformer, when *TxEnable* is OFF and the frontend enters in RX-mode. Transistors M3-4 use thick gate oxide ($t_{ox} = 5\,$nm) with the associated higher breakdown voltage as they have to withstand peak drain voltages above $2 \cdot V_{DD}$ due to the class-E operation.

The receiver does not employ a frontend LNA and, therefore, the transformer is passively connected to the down-converter through the NMOS RX switches M5-6. Similar as for M3-4, these switches are implemented with thick oxide transistors. Quadrature down-conversion is performed by two passive mixers (M7-M10 in Fig. 4.33), which basically steer the incoming RF current into the baseband load. No DC current is carried by the mixers because the RX-switches are AC-coupled to the transformer. Passive mixers offer high linearity, contribute no flicker noise [108], and provide a compact solution due to the increased internal impedance level. This is because higher on-resistances can be tolerated and, hence, smaller mixing transistors M7-10 can be used ($w = 8\,\mu$m and $l = 130\,$nm in this design). Additionally, by using small gate areas, the passive mixers add little capacitances to the QVCO tank; parasitics that can be readily tuned out by slightly modifying the inductors of the oscillator.

The first active stage in the RX-path is the baseband transimpedance amplifier (TIA) at the output of the quadrature mixer. It is an opamp-based current buffer which ideally creates a virtual, differential ground at its input [83]. Mixer output currents are passed through the feedback network $Z_f = R_f || C_f$ and converted into a differential output voltage. This feedback network also defines the DC transimpedance of the TIA and implements the first channel filtering pole of the receiver at approximately 1 MHz.

Figure 4.34 shows the schematic of the differential operational amplifier used in the TIA. It consists of two stages. The first stage is a differential PMOS input pair, MP1-2, actively loaded by two common-source NMOS transistors, MN1-2. It obtains a differential voltage gain of about 30 dB. The second stage is a voltage buffer used to drive the load impedance of the amplifier, essentially dominated by the feedback network Z_f. The buffer uses two source followers MP3-4 and MN3-4 in cascade, so that the input and output terminals of the operational amplifier can share the same DC voltage level and, hence, no DC current flows through the feedback network of the TIA. The input/output common-mode voltage of the amplifier is set to 200 mV by means of a common-mode feedback network (not shown in Fig. 4.34). Such a low voltage value provides enough overdrive for the switching of the passive mixers and the pass transistors M5-6.

In order to prevent the output of the TIA from saturating at high RF input levels, an attenuator is included in the RX-path. It is implemented by an array of four minimum-length NMOS transistors (M11 in Fig. 4.33) connected at the antenna port. This also precludes strong incoming RF signals from entering the transformer and creating a magnetic field that could disturb the QVCO. Transistors M11 are binary weighted and sum a total width of 384 μm. The attenuation is set by a thermometer-code control word *att*[3:0] to values between 7 and 20 dB in steps of about 4 dB. The control of

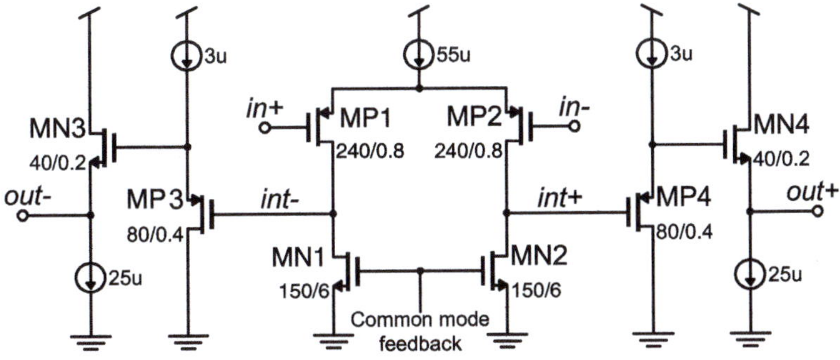

Fig. 4.34 Schematic of the operational amplifier employed in the TIA

the attenuator and the baseband PGA is done off-chip in an FPGA based on the outputs of the overflow detectors placed at corresponding points in the baseband amplification chain (see Fig. 4.1).

4.2.3 Experimental Results

The transceiver frontend has been integrated in the same 130 nm CMOS technology as before using metal-insulator-metal (MIM) capacitors (Fig. 4.35). All measurements have been carried out with the test chip 3 (Fig. 4.20), assembled again in a QFN-36 package and soldered onto a test board. In order to perform RF measurements, a surface-mount balun with 2:1 impedance ratio [1] has been used to convert the

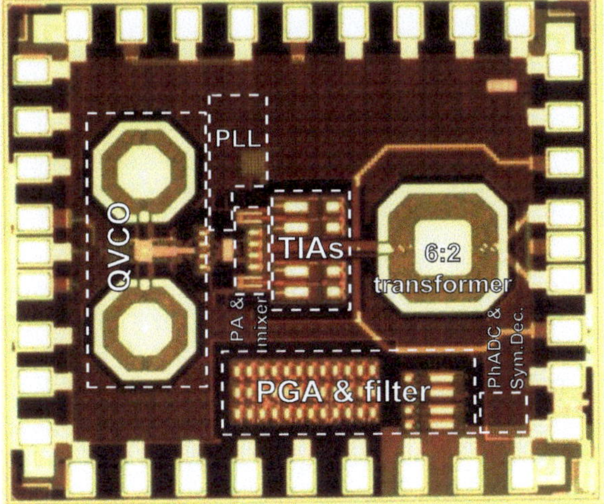

Fig. 4.35 Microphotograph of the complete transceiver (chip 3). The die size is 2.1 mm² (1.6 by 1.3 mm) including the pads

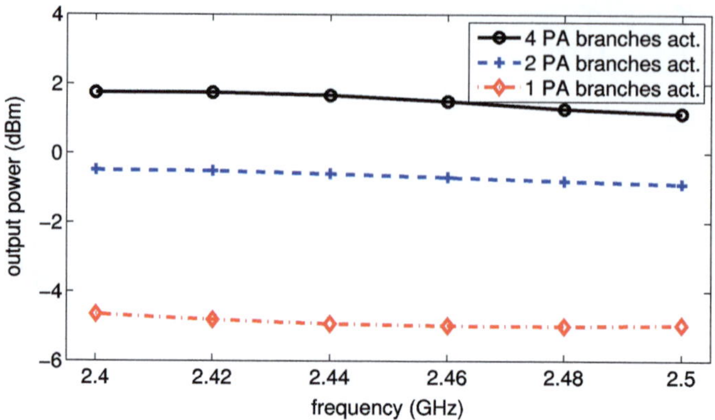

Fig. 4.36 Output spectrum of the transmitter at maximum power level

differential $100\,\Omega$ antenna interface of the chip into the single-ended $50\,\Omega$ standard impedance of the laboratory equipment. A supply voltage of $1.0\,$V has been used in all experiments.

The output power P_{out} of the transmitter has been measured by using a power meter and applying de-embedding techniques to remove the losses of the external balun. As shown in Fig. 4.36, the output power at the maximum gain setting, i.e., with the four PA branches enabled, varies by only $0.4\,$dB along the BLE frequency band and peaks at $2.42\,$GHz with $1.6\,$dBm. The output stage of the PA consumes $4.8\,$mA and so achieves a power efficiency of $30.1\,\%$. If the power consumption of the PA drivers is taken into account, the supply current rises to $5.44\,$mA and the total efficiency is $26.6\,\%$. To reduce the power consumption, the PA may be operated with only 1 branch activated, i.e., in power back-off mode. In this case, the output power is $-5\,$dBm and the current consumption $2.46\,$mA (incl. the driver), leading to a reduced total efficiency of $12.9\,\%$.

Figure 4.37 illustrates the input matching of the transceiver in RX-mode, measured through the balun with the HP8719D network analyzer. In this setup, the frequency synthesizer is set to $2.44\,$GHz which causes a small discontinuity of the return loss at that frequency. Nevertheless, a broadband matching is obtained across the ISM band of 2400–$2483.5\,$MHz, i.e., a return loss below $-15\,$dB is measured throughout this bandwidth.

Figure 4.38 shows the measured receiver linearity for the maximum attenuation setting of $19.6\,$dB. The 1-dB compression point (*IP1dB*) is $-14.4\,$dBm and the third-order intermodulation intercept point (*IIP3*) is $-2.9\,$dBm, both referred to the antenna input of the chip.

The noise performance of the receiver frontend has been evaluated from the antenna port to the TIA outputs. Figure 4.39 shows that the noise figure (NF) of the receiver remains between 16.0 and $16.6\,$dB within the ISM-band. This relatively poor NF is mainly caused by the passive mixer, which features a negative conversion gain of approximately $-11\,$dB, according to simulations. In return, the power

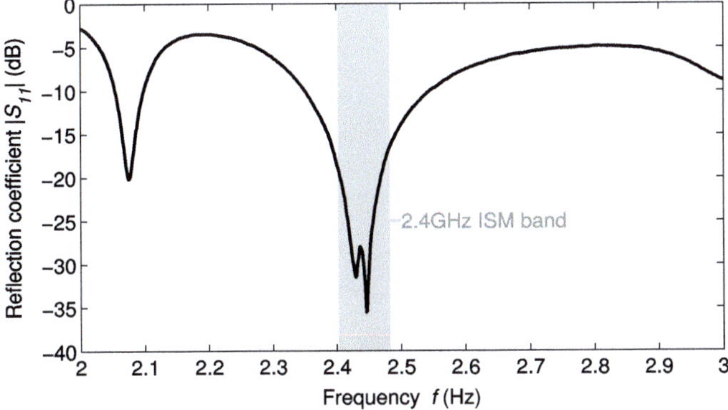

Fig. 4.37 Return loss $|S_{11}|$ in RX-mode

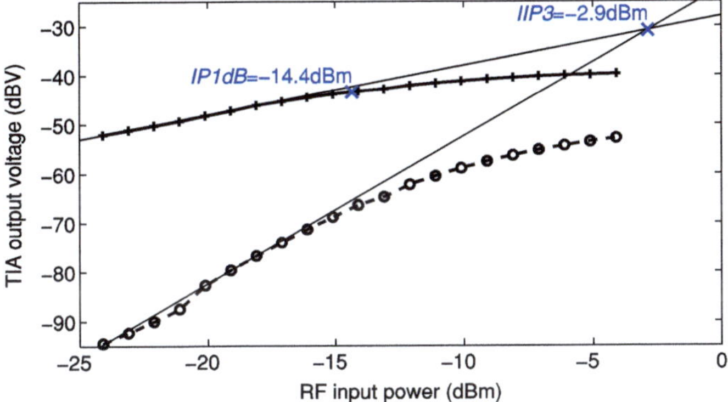

Fig. 4.38 Measured RX linearity with attenuator activated

consumption of the receiver frontend is only 230 μW, which corresponds to the dissipation of the TIAs.

4.2.4 Conclusion

A low-power zero-IF transceiver frontend architecture has been implemented, which employs a step-up transformer to up-convert the internal RF impedance. This allows for a class-E power amplifier with a low output power of 1.6 dBm that operates efficiently from a 1.0 V-supply. The measured efficiency of the output stage alone is 30.1 %, while the complete PA with drivers achieves an efficiency of 26.6 %. The RF section of the receiver frontend is completely passive and also takes advantage of the impedance transformation because it provides passive voltage gain. The first stage of active amplification in the receiver are the transimpedance amplifiers (TIAs) at the output of the passive mixers, which consume together 230 μA from a 1 V-supply.

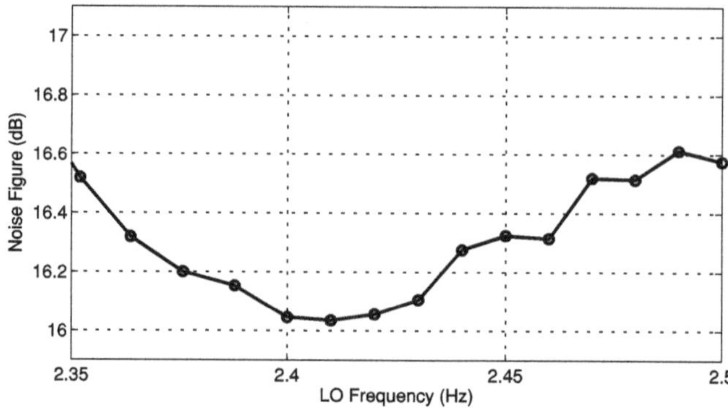

Fig. 4.39 Noise figure of the RX frontend, measured at the TIA output ($f_{in} = f_{LO} + 200\,\text{kHz}$, $P_{in} = -70\,\text{dBm}$)

With this approach, the power consumption is minimized while sacrificing noise performance. Still, the average noise figure of 16.2 dB is sufficiently low to allow for a BLE sensitivity better than $-80\,\text{dBm}$, as long as a GFSK-demodulator is employed that requires a bit-energy to noise-floor ratio (E_b/N_0) of 17.8 dB or less. As will be shown in the following section, this can be achieved easily.

4.3 Phase-Domain Zero-IF Demodulator

As explained in the previous chapter (Sect. 3.4), a low-complexity demodulator based on a phase-domain ADC (Ph-ADC) has been selected for the proposed ultra low power transceiver. This architecture has the advantage that it requires few quantization steps, i.e., a 4-bit Ph-ADC is sufficient, and that performs well over a wide dynamic range. Therefore, the gain equalization can be done with coarse steps, facilitating a low power implementation. In this section, the implementation of the phase-domain zero-IF demodulator will be detailed.

4.3.1 Architecture

The block diagram of the proposed zero-IF GFSK demodulator is shown Fig. 4.40. The I- and Q-signals are first filtered and equalized with a two-stage Programmable Gain Amplifier (PGA) over a dynamic range of more than 50 dB. Due to the fact that the subsequent Ph-ADC only evaluates the phase information, the PGA only implements coarse gain steps of 6 dB. In order to allow for an external gain control, the output voltages of the PGA stages are monitored by overflow detectors.

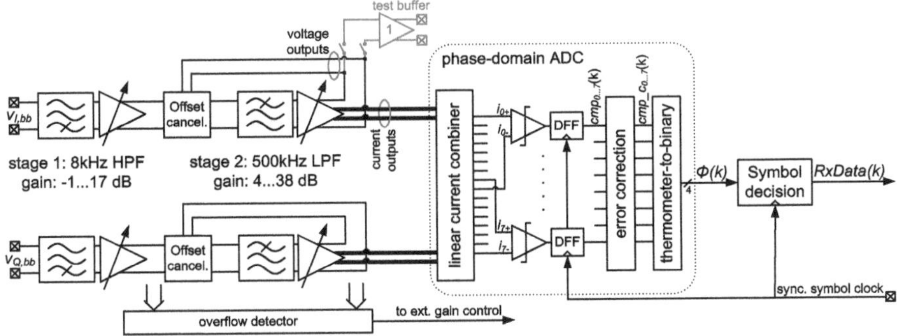

Fig. 4.40 Overview of the presented demodulator

The demodulator is tailored to the recently published Bluetooth low energy (BLE) standard which defines a GFSK input signal with a modulation index of $h = 0.5$ [3]. This means that the complex phasor rotates by $\pm 90°$ ($\pm h\pi$) per symbol. Therefore, a 4-bit Ph-ADC with a phase resolution of $22.5°$ ($\pi/8$) is sufficient to detect the phase rotation. With 4-bit quantization, the demodulator performance is degraded by only 0.5 dB with respect to a theoretic phase-domain demodulator without quantization. A lower 3-bit quantization would degrade the performance by about 4 dB, whereas for 5-bit or higher no substantial improvement is observed. The analog part of the Ph-ADC, namely the linear combiner for phase rotations and the comparators, operates completely in current-domain. This new method not only renders resistors in the linear combiner unnecessary but also directly taps the required currents from the output branch of the PGA and so skips any buffering stage. Both of these aspects are essential to achieve low power consumption. In the digital post-processing of the Ph-ADC, the comparator outputs are sampled, passed through an error correction circuit, and mapped into a binary-coded 4-bit phase $\phi(k)$. Finally, the symbol decision block compensates carrier frequency offsets and outputs the demodulated data.

Considering the BLE symbol rate of $f_{sym} = 1\,\text{MHz}$, the received GFSK signal occupies a 3dB-bandwidth of 800 kHz, which is ideally centered around DC after zero-IF down-conversion. In practice, this signal is shifted by the carrier frequency offset between transmitter and receiver, which may be up to $\pm 150\,\text{kHz}$ in a BLE system. In order to keep the noise bandwidth of the demodulator as low as possible, a low-pass filter (LPF) cut-off frequency of 500 kHz has been chosen. This is sufficient to cope with carrier frequency offsets as will be shown in the experimental results (Sect. 4.3.5). To meet the BLE blocking specifications, the proposed demodulator requires only a third-order LPF, taking into account that the preceding receiver frontend already provides a first-order filtering.

In other respects, flicker noise and DC offsets are always a concern in a zero-IF architecture because they fall into the bandwidth of the received signal. Unavoidably, removing these disturbances with a high-pass filter (HPF) will also cut a portion of the signal content. Therefore, care must be taken on the selection of the HPF cut-off frequency, i.e., it must be high enough to reject flicker noise and obtain a fast system response, and low enough to maintain signal integrity. In order to evaluate this tradeoff, the performance of the zero-IF GFSK demodulator has been simulated with behavioral models assuming varying HPF cut-off frequencies f_{HPF}. The bit-error-rate (BER) simulations, shown in Fig. 4.41a, are based on an ideal 4-bit Ph-ADC based demodulator, assuming a flicker noise corner of 150 kHz as expected at the TIA output of the frontend. Considering the typical BER benchmark of 0.1 %, the lowest required ratio of energy-per-bit to white noise density (E_b/N_0) is obtained with $f_{HPF} \approx 2$ kHz, as shown in Fig.4.41b. This optimum rises to about 4 kHz for a BER of 1 %. On the other hand, the 10 % settling time $t_{settle,10\%}$ of the HPF's step response is inversely proportional to the cutoff frequency, i.e., $t_{settle,10\%} = 0.366/f_{HPF}$. Therefore, in this implementation the demodulator uses a HPF cut-off frequency of 8 kHz and so keeps $t_{settle,10\%}$ below 50 µs at the expense of an acceptable performance degradation of 0.2–0.5 dB. It is worth observing the minimum interval between two BLE data packets is specified as 150 µs, what gives room enough for settling.

4.3.2 PGA and Channel Filter

In order to cover a dynamic range of more than 50 dB, the programmable gain amplifier (PGA) employs two stages. The first stage, with a moderate amplification of up to 17 dB, is optimized for low input noise and sets the overall HPF cutoff

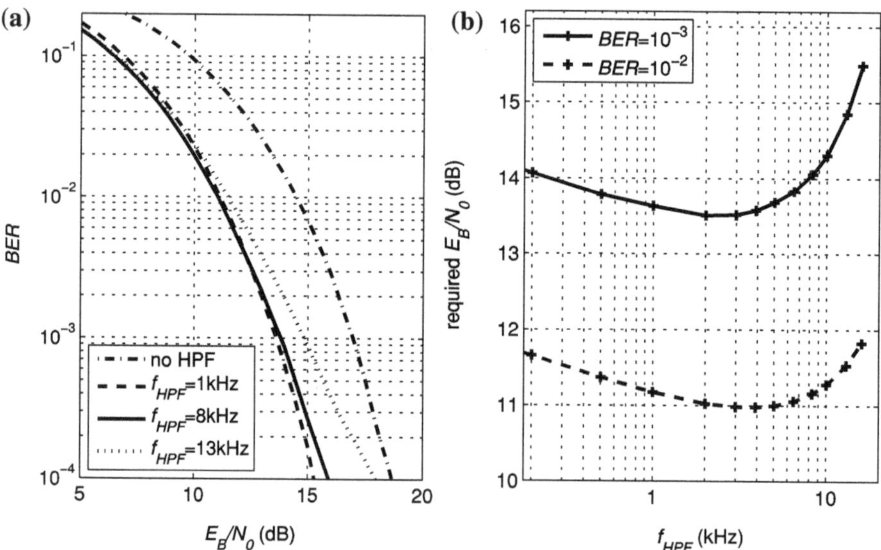

Fig. 4.41 Simulation of the influence of the HPF corner frequency f_{HPF} using an ideal 4-bit Ph-ADC based demodulator with a flicker noise corner of 150 kHz. **a** Bit-error-rate (BER) versus ratio of energy-per-bit to white noise density (E_b/N_0). **b** Required E_b/N_0 to achieve a given BER

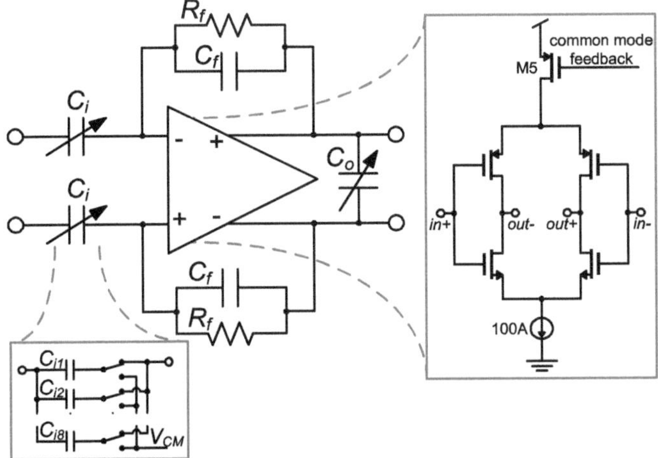

Fig. 4.42 Schematic of the first PGA stage

frequency. The second stage provides a large amplification of up to 38 dB and defines
the overall LPF cut-off frequency. DC offsets are removed by means of a cancelation
loop around the second stage.

The first stage of the PGA, shown in Fig. 4.42, employs capacitive feedback
to set the gain according to the capacitance ratio C_i/C_f. Programmability with
6dB-steps is obtained by implementing the input capacitance C_i as an array of eight
identical metal-insulator-metal (MIM) capacitors $C_{i1...8} = C_f = 2.4\,\text{pF}$. The HPF
time constant is set by the feedback network $R_f \| C_f$ and hence does not depend on
the gain setting. The high-resistive Poly resistor R_f with a nominal value of $8.3\,\text{M}\Omega$
can be calibrated with a 2-bit control word to compensate process variations.

In order to achieve both low input noise and low power consumption, the active
part of the first stage is implemented as a complementary transconductance stage,
shown in the inset of Fig. 4.42. In this configuration, the NMOS and PMOS transistors
reuse the same bias current and so reduce the input referred noise by a factor of two
[102]. The load capacitance C_O of the amplifier is implemented as a tunable MOS
capacitor array. This array is internally adjusted to counteract the dependency of the
output pole with the gain setting and, hence, keep the low-pass corner frequency of
the stage at roughly 1.5 MHz, regardless of the C_i value.

The dominant poles of the PGA are set by the second stage which employs a
Sallen-Key structure [127]. In order to maintain a constant transfer function, the
programmable forward gain A_2 is compensated with an attenuator F_2 in the Sallen-
Key feedback path, as illustrated in Fig. 4.43a. Using this approach, the passives R_1,
R_2, C_1, and C_2 do not have to be changed for the different gain settings.

The fully differential implementation of the second PGA stage is shown in
Fig. 4.43b. The amplifier employs a g_m-boosted differential pair M1/M2 with source
degeneration [30]. In this configuration the DC gain is defined by the ratio of the

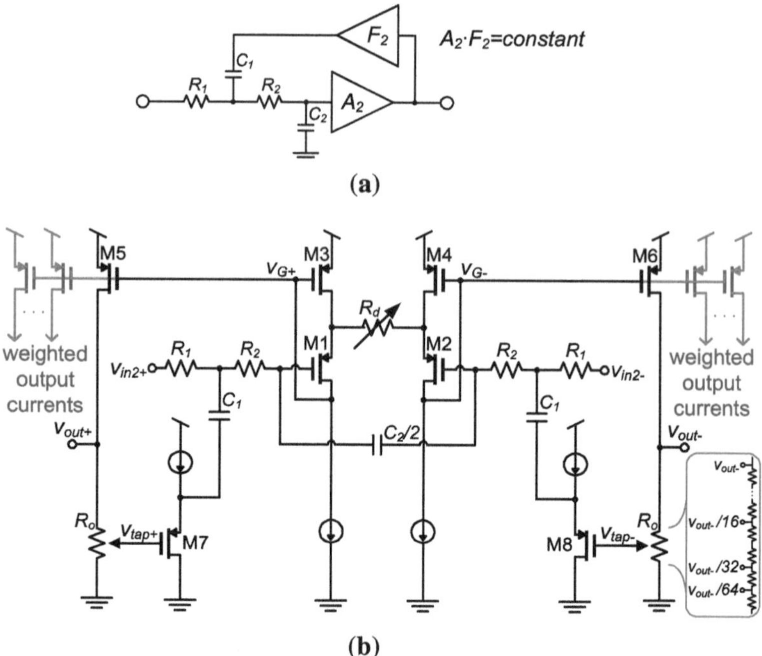

Fig. 4.43 Second PGA stage. **a** Programmable Sallen-Key structure. **b** Schematic of the implementation

output resistor R_o to the degeneration resistor R_d. Due to the scaled current mirrors M3/M5 and M4/M6 (ratio 4:1), the currents in the output branches are reduced and the nominal gain of the amplifier calculates as

$$A_2 = \frac{R_o}{4R_d/2} = \frac{R_o}{2R_d}.$$

(4.20)

The gain is programmed by adjusting the source degeneration resistor R_d. As this resistor is connected between internal nodes of the gain stage, the poles of the transfer function are kept essentially constant. Note that the output currents required by the subsequent Ph-ADC are readily available from this structure by tapping the current mirrors M3/M5 and M4/M6.

In order to obtain a large programming range of nominally 36 dB the degeneration resistor has to be adjusted over almost two orders of magnitude. This is achieved by implementing R_d as eight unit resistors of 12.5 kΩ which can be connected in different serial and parallel configurations. Figure 4.44 depicts the detailed setup of this resistor array and Table 4.5 shows the states of the switches that result in a resistance range for R_d from 3.125 to 100 kΩ in 6 dB steps. Note that there are always at least four unit resistors used in order to maintain good matching.

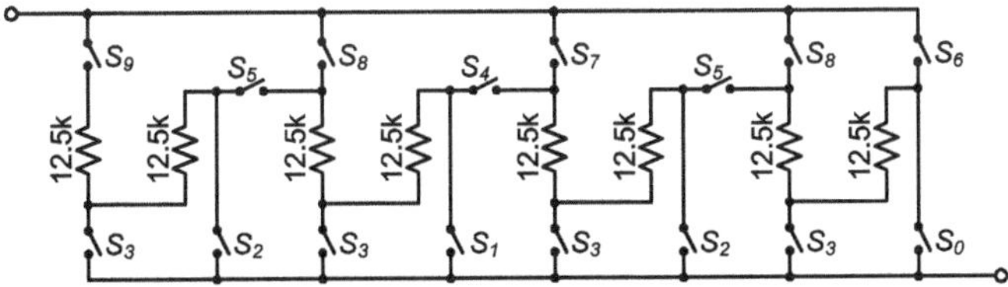

Fig. 4.44 Implementation of R_d and switch setup for the possible gain settings

Table 4.5 Switch setup for degeneration resistor R_d

A_2 (dB)	$S_{9:0}$ [a]	R_d (kΩ)
6	1000 11 0001	100
12	1000 10 0010	50
18	1010 10 0011	25
24	1010 00 0100	12.5
30	1110 00 0111	6.25
36	1110 00 1000	3.125
42	1111 11 1000	1.5625

[a] A shorted switch is represented by a 1

The output resistor R_o of the second PGA stage has a constant value of $400\,k\Omega$ and it is implemented as a series connection of 64 equally sized resistors of $6.25\,k\Omega$, as shown in the inset of Fig. 4.43b. This structure allows to control the attenuation of the Sallen-Key feedback by properly tapping the internal nodes of the resistive ladder as a function of the gain setting of the PGA stage. For the smallest gain setting, tapping is done at the top of the ladder so that $v_{\text{tap}\pm} = v_{\text{out}\pm}$, whereas, for the largest gain, tapping is done at the bottom of the ladder so that $v_{\text{tap}\pm} = v_{\text{out}\pm}/64$. The tapped voltage is buffered by the source follower M7/M8 and applied to the feedback capacitor of the Sallen-Key filter C_1.

Similar as for the first stage, the two high-resistive polyresistors R_1 and R_2, which define the cut-off frequency of the Sallen-Key filter, can be jointly calibrated with a single 2-bit control word to compensate for process variations. This is possible, since C_1 and C_2 are both MIM-capacitors and so the two decisive time constants $R_1 C_1$ and $R_2 C_2$ are equally affected by the process. The 2-bit calibration effectively reduces the process-induced error of the cut-off frequency from $\pm 16\,\%$ to about $\pm 4\,\%$, which is sufficient to meet the blocking requirements. No automatic self-tuning mechanism is provided in the current implementation.

In order to prevent the output of the PGA from saturation due to mismatch, an offset cancelation loop is necessary. In the proposed PGA, offset is compensated at the interface of the two stages with the circuit shown in Fig. 4.45. The NMOS source follower pair M9/M10 acts as a current-controlled unity gain level shifter, which connects the output of the first PGA stage to the input of the second stage. Offset compensation is accomplished by unbalancing the bias currents of the follower pair based on the difference between the PGA output voltages, which is integrated in the

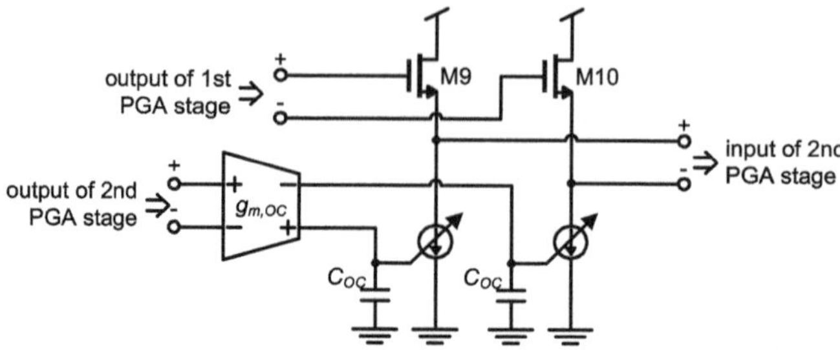

Fig. 4.45 Offset compensation circuit

capacitors C_{OC} through the transconductance $g_{m,OC}$. The transconductor is realized as a differential PMOS pair with common mode feedback that operates in weak inversion in order to obtain a large time constant of 150 μs for the integration. The offset cancelation loop adds a zero to the PGA transfer function, which is nondominant for all gain settings due to the slow integration.

4.3.3 Phase-Domain ADC

4.3.3.1 Design of the Analog Frontend

For an N-ary Ph-ADC, the IQ-plane is split up by defining $N/2$ thresholds with an angle of $2\pi/N$ between consecutive quantization intervals, as illustrated in Fig. 4.46. For the proposed 4-bit Ph-ADC ($N = 2^4$), the pth threshold Th_p is characterized by

$$-I \cdot \sin \frac{p\pi}{8} + Q \cdot \cos \frac{p\pi}{8} = 0, \qquad p = 0, 1, \ldots, 7 \qquad (4.21)$$

with I and Q being the in-phase and quadrature component of the input signal, respectively. Therefore, the operation principle of the Ph-ADC simply relies on detecting the zero crossings of eight linear combinations of the I and Q signal components and mapping the result to a 4-bit result vector.

Conventionally, the linear combiner has been implemented using a resistive bridge that converts input currents to phase-shifted output voltages [21, 37, 67, 80]. Figure 4.47a shows a resistor-based linear combiner that directly uses the differential I and Q current inputs and generates the required $\pi/8$ phase intervals by making $R_A/R_B = \sqrt{2}$ [67, 80]. A bridge with equal resistances is also possible but it needs additional input currents, i.e., the sum and difference of the I and Q currents scaled with a factor $\alpha = 1/\sqrt{2}$, as shown in Fig. 4.47b [21, 37]. However, both resistive networks have two main drawbacks. First, they lead to an inevitable tradeoff between area occupation and power consumption, i.e., in order to obtain reasonable output voltages either large resistors or large input currents have to be used. This tradeoff will

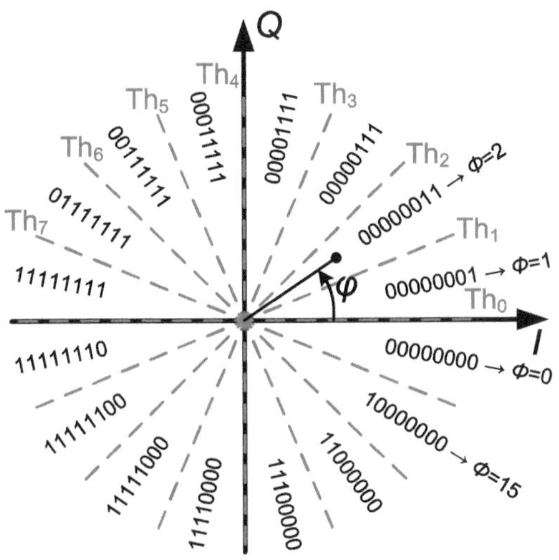

Fig. 4.46 Quantization of the complex IQ-plane with a phase-domain ADC. The 8-bit words represent the comparator outputs $cmp_{7...0}$ corresponding to the thresholds Th_p

be detailed shortly in Sect. 4.3.3.2 by comparing it to the proposed solution. Second, the phase-rotated outputs of the resistive networks exhibit different amplitudes by construction. If we consider a rotating phasor with constant magnitude, the bridge in Fig. 4.47a leads to a maximum amplitude imbalance of $-3\,\mathrm{dB}$ ($\frac{\hat{v}_{2\pm}}{\hat{v}_{0+}} = \cos\frac{\pi}{4}$) while the bridge in Fig. 4.47b yields a maximum imbalance of $-0.7\,\mathrm{dB}$ ($\frac{\hat{v}_{1\pm}}{\hat{v}_{0+}} = \cos\frac{\pi}{8}$). Although the phase rotated outputs will be hard-limited in the subsequent stage of the Ph-ADC, different amplitudes, and hence different slopes at the zero-crossings, may cause an imbalanced dynamic behavior and eventually increase the dynamic nonlinearity of the Ph-ADC. Therefore, both angular and amplitude error of the linear combiner should be minimized.

In the proposed Ph-ADC, the generation of the linear combinations is performed in the current domain without any resistor. The output currents of the PGA are weighted with integer multiplicity-factors m and combined as shown in Fig. 4.48. Hence, the scaling factors in (4.21) are effectively approximated by rational numbers as follows

$$\sin\frac{\pi}{8} = \cos\frac{3\pi}{8} \approx \frac{5}{13} \tag{4.22}$$

$$\sin\frac{\pi}{4} = \cos\frac{\pi}{4} \approx \frac{9}{13} \tag{4.23}$$

$$\sin\frac{3\pi}{8} = \cos\frac{\pi}{8} \approx \frac{12}{13} \tag{4.24}$$

$$\sin\frac{\pi}{2} = \cos 0 = \frac{13}{13}. \tag{4.25}$$

The common denominator 13 has been chosen because it produces only a small angular error of $0.12°$ ($\arctan\frac{5}{12} - \frac{\pi}{8}$) in the differential output currents $i_{1,3,5,7}$.

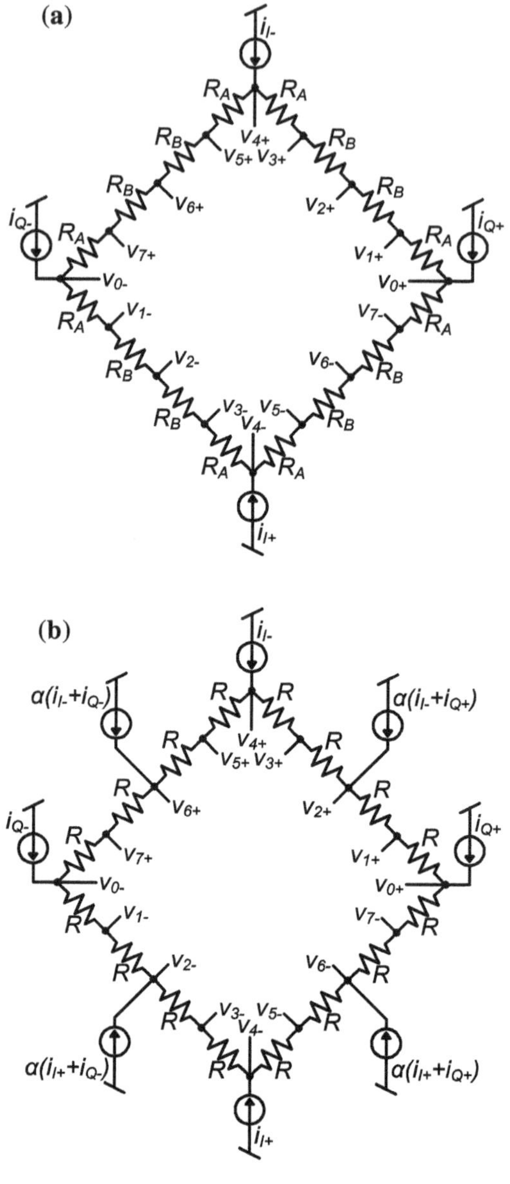

Fig. 4.47 Conventional techniques for phase-rotation with resistive bridges (common mode resistors from input nodes to ground not shown). **a** With two differential input currents. **b** With four differential input currents

Moreover, the amplitudes of these outputs exactly coincide with the amplitudes of $i_{0,4}$ since 5, 12, and 13 form a Pythagorean triple ($5^2 + 12^2 = 13^2$). Only the differential output currents $i_{2,6}$ carry a small amplitude error of $-0.18\,\mathrm{dB}$ ($\frac{\hat{i}_{2\pm}}{\hat{i}_{0+}} = \sqrt{2}\frac{9}{13}$).

The eight differential outputs of the linear combiner are passed to corresponding differential current comparators, which are realized according to the schematic of

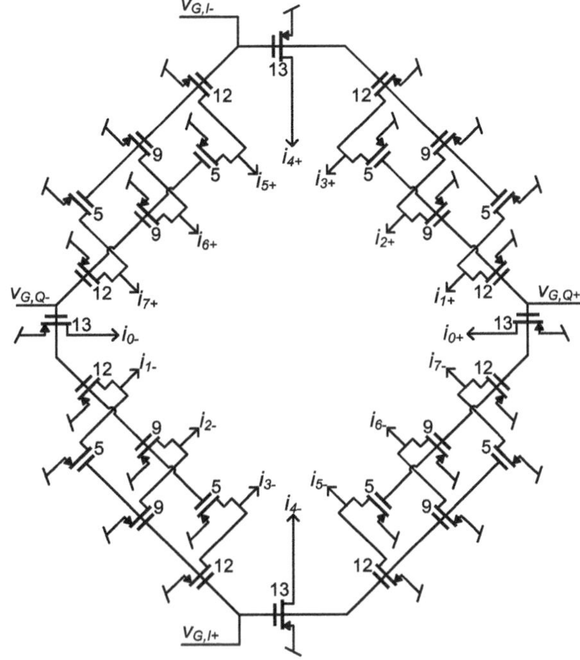

Fig. 4.48 Linear current combiner based on current source, with multiplicity factors m shown next to the respective transistor (effective gate width $w = m \cdot 0.5\,\mu m$, gate length $l = 4\,\mu m$)

Fig. 4.49. The current mirror MN2/MN3 converts the differential input to a single-ended current Δi_p. This current Δi_p is then compared to zero by a current steering comparator which is known for its high speed and virtually zero-offset operation [112]. It consists of a simple operational transconductance amplifier (OTA) with a differential PMOS input and a nonlinear feedback network formed by the transistors MN1 and MP1. This nonlinear feedback serves two purposes. First, it pulls v_{p+} to the same level as v_{p-}, which is set by the gate voltage of the current mirror MN2/MN3. This ensures not only equal operating conditions at the current mirror MN2/MN3 but also symmetric loading of the outputs of the linear combiner, i.e., the two outputs to be compared see the same load voltage. Second, at the quiescent point ($\Delta i_p = 0$, $v_{p+} = v_{p-}$) both feedback transistors have a gate-source voltage of zero which leads to a transimpedance of the current comparator in the $G\Omega$-range.

4.3.3.2 Phase Error Analysis and Comparison

In order to illustrate the advantages of the proposed current-domain Ph-ADC with respect to the resistive network based Ph-ADC, a simple model has been developed to disclose the tradeoff between current consumption and accuracy for both linear combiners.

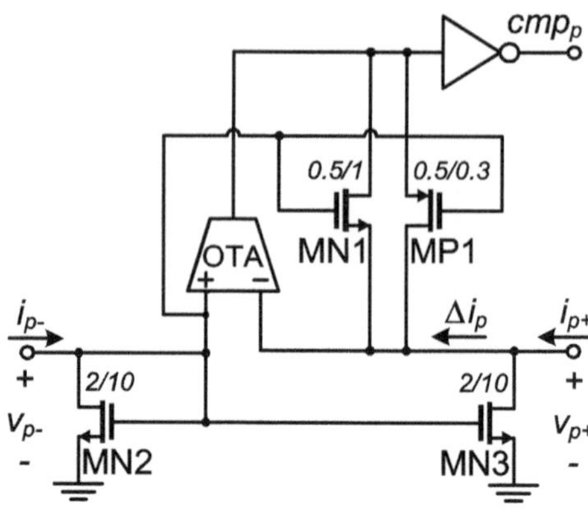

Fig. 4.49 Differential current comparator with gate dimensions (w/l) shown in μm

Let us first consider the current-domain Ph-ADC and assume there exists an offset current ΔI_C between the positive and negative branches of the comparator of Fig. 4.49. Given a rotating phasor with peak amplitude $I_{C,\mathrm{PK}}$ and angular frequency ω, the branch currents to compare $i_{p\pm}$ are sinusoidal waves as shown in Fig. 4.50. To avoid negative currents, a common-mode current $I_{C,\mathrm{CM}} = I_{C,\mathrm{PK}}/\alpha$ is assumed, where α is slightly less than 1. The phase errorwhich results from the difference between the ideal and the offset-shifted crossing of $i_{p\pm}$ can be calculated as

$$\Delta\varphi_C = \arcsin\left(\frac{\Delta I_C}{2I_{C,\mathrm{PK}}}\right) \approx \frac{\Delta I_C}{2\alpha I_{C,\mathrm{CM}}} \tag{4.26}$$

where it is assumed that $\Delta I_C \ll 2I_{C,\mathrm{PK}}$ as occurs in practical situations. This offset current can be related to the mismatch of the mirror MN2/MN3 and the mismatch of the PMOS current sources of the combiner by means of the well-known Pelgrom's model [97] as

$$\sigma^2(\Delta I_C) = I_{C,\mathrm{CM}}^2 \cdot \left(\frac{A_{\beta,P}^2}{WL_P} + \frac{A_{VT0,P}^2}{V_{\mathrm{od},P}^2 \cdot WL_P} + \frac{A_{\beta,N}^2}{WL_N} + \frac{A_{VT0,N}^2}{V_{\mathrm{od},N}^2 \cdot WL_N}\right) \tag{4.27}$$

where A_β and A_{VT0} are the technological parameters for current and threshold voltage mismatch, respectively, V_{od} is the overdrive voltage and WL the gate area of the transistors (the sub-indices P and N indicate the PMOS and NMOS section, respectively). Assuming these transistors are biased in strong inversion and that threshold voltage mismatch dominates [97], the offset current is found to be proportional to $\sqrt{I_{C,CM}}$ and, thereafter, $\Delta\varphi_C \propto 1/\sqrt{I_{C,CM}}$. Note that the offset of the OTA in Fig. 4.49 can be neglected due to integrating nature of the current comparator [112].

Similarly, the phase error $\Delta\varphi_R$ of the resistor-based combiner in Fig. 4.47a can be estimated by taking into account the current mismatch of the driving current sources, the mismatch between the resistors, and the offset of the voltage comparator. All these error sources can be referred back to an input offset current of the driving

Fig. 4.50 Effect of offset on detected zero crossing

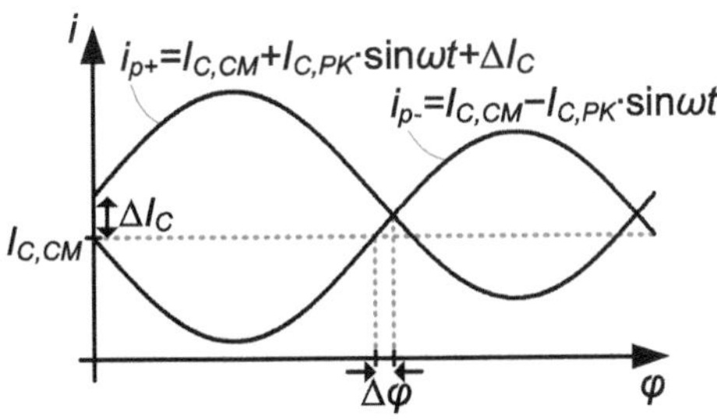

sources, ΔI_R, as

$$\sigma^2(\Delta I_R) = I_{R,CM}^2 \cdot \left(\frac{A_{\beta,Src}^2}{WL_{Src}} + \frac{A_{VT0,Src}^2}{V_{od,Src}^2 \cdot WL_{Src}} + \frac{A_R^2}{WL_R} \right) + \frac{A_{VT0,Cmp}^2}{R_{br}^2 \cdot WL_{Cmp}} \quad (4.28)$$

where the same nomenclature as before applies and the subindices Src, R and Cmp refer to the driving sources, branch resistors $R_{br} = 2(R_A + R_B)$ and comparator, respectively. Accordingly, a similar expression as (4.26) can be likewise applied to determine $\Delta\varphi_R$. Note from (4.28) that whereas the mismatch of the current sources has similar impact as in (4.27), the comparator offset voltage $\Delta V_{Cmp} = A_{VT0,Cmp}^2/WL_{Cmp}$ gives rise to a phase error which is inversely proportional to the common-mode input current $I_{R,CM}$ of the driving sources, i.e., $\Delta\varphi_R(\Delta V_{Cmp}) \propto \Delta V_{Cmp}/(I_{R,CM} \cdot R_{br})$. Additionally, the resistor mismatch also affects the phase error with an inverse square-root relation, $\Delta\varphi_R(R_{br}) \propto 1/\sqrt{R_{br}}$, provided the resistors have a constant width.

Figure 4.51 shows the estimated standard deviation of the phase error $\sigma(\Delta\varphi)$ for both alternatives in terms of the total current consumption of the linear combiner, I_{TOT}, and the branch resistance, R_{br}. In both cases, the matching parameters are taken from the selected 130 nm-CMOS technology, transistors are operated in strong inversion and $\alpha = 0.8$. The transistor dimensions of the current-domain method are as shown in Figs. 4.48 and 4.49, while for the resistor-based approach the same PMOS current sources are assumed ($WL_{Src} = 26\,\mu m^2$) and the comparators have a gate area of $WL_{Cmp} = 20\,\mu m^2$, which yields $\sigma(\Delta V_{Cmp}) = 0.7\,mV$ in this technology. In the case of the resistor-based combiner, either small currents or small resistances lead to high phase errors as a result of the comparator offset and the resistance mismatch, respectively. Contrarily, the phase error improves for large resistances and biasing currents. In this part of the design space, the driving sources dominate the phase error. In the case of the current-domain Ph-ADC, the standard deviation of the phase error, which obviously does not vary with the branch resistance, exhibits less degradation at low biasing currents because of the smoother dependence of $\Delta\varphi_C$ on the DC current. Figure 4.51 also shows the projection of the intersection of the two surfaces (dashed gray line) to bring out the regions in which one approach outperforms the other. Taking also into account the limit imposed by the rail-to-rail voltage range, it can be observed that the resistive network based Ph-ADC only outperforms the accuracy

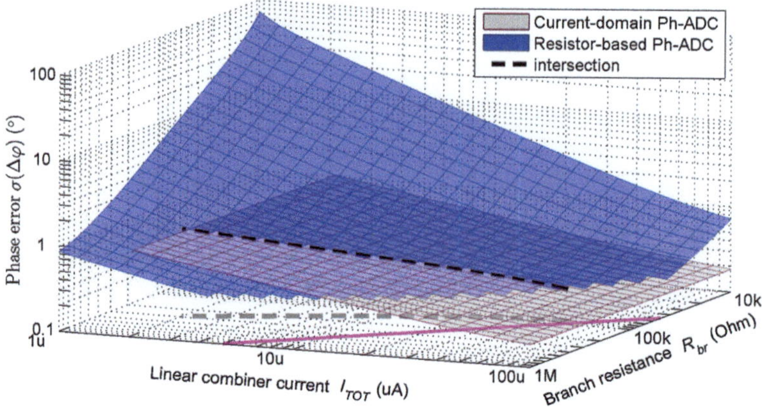

Fig. 4.51 Phase error comparison of current-domain method versus resistor-based approach according to Fig. 4.47a for varying the design variables. The *dashed gray line* projects the intersection of the two surfaces down into the $I_{LC} - R_{br}$-plane and the *solid line* displays the upper limit of the $I_{LC} R_{br}$ product for the resistor-based method imposed by a supply voltage of 1 V

of the current-domain alternative in a small region of the design space. However, in such region large resistances are needed, leading to high area occupation. Moreover, the phase error is only up to $\sqrt{2}$-times lower than for the current-domain approach in this example, since we assumed equal PMOS current sources and the current mirror contributes about the same offset as these sources. On the other hand, for small resistances and currents, i.e., low area occupation and power consumption, the performance advantage of the current-domain alternative is much more significant.

The basic trend highlighted by this simple analysis is also confirmed with experimental results. Considering all mismatch sources of the second PGA stage and the analog part of the Ph-ADC, a phase-error of $\sigma(\Delta\varphi) = 1.2°$ was obtained from Monte-Carlo analysis for the implemented current-domain Ph-ADC. The complete 4-bit Ph-ADC consumes about 25 μA in simulation, 10.8 μA thereof in the linear combiner, and occupies an area of 0.015 mm^2 in a 130 nm-CMOS technology. Compared to the resistor-based 4-bit Ph-ADC in [21], which is the only publication giving details on current consumption and area, the proposed current-domain solution consumes about one order of magnitude less current (25 μA w.r.t. 290 μA) and occupies one-third of the area (0.015 mm^2 w.r.t. 0.044 mm^2), although the latter is partially due to the larger 180 nm-CMOS technology used in [21].

4.3.3.3 Digital Post-Processing

In the digital post-processing, the thermometer coded comparator outputs $cmp_{0...7}$ are first sampled at the symbol rate with D-flipflops (DFFs) and the sampled data, $cmp_{0...7}(k)$, is transformed to the binary-coded 4-bit output vector of the Ph-ADC

Fig. 4.52 Majority decoder

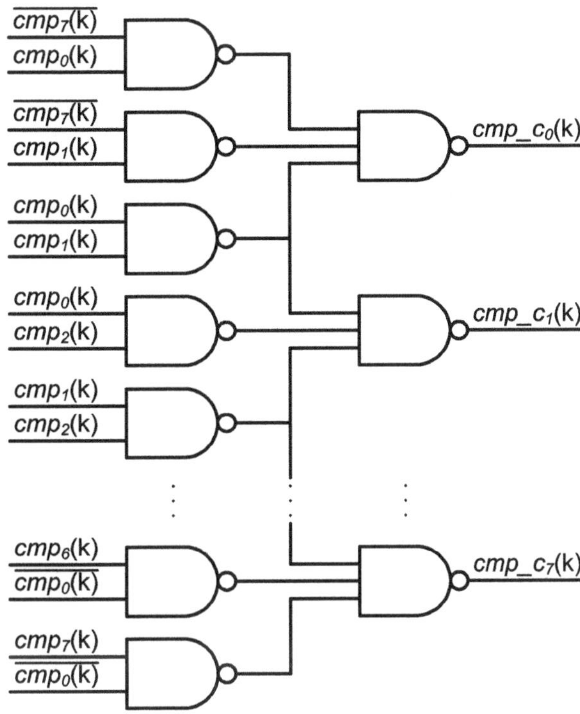

$\Phi(k)$. In order to correct noise-induced bit errors on $cmp_{0...7}(k)$ a majority decoder (Fig. 4.52) precedes the code conversion [78]. The majority decoder uses the input bit together with its two neighbors and so repairs all single-bit errors in the thermometer code. These errors, which are often referred to as "bubbles", are either corrected (e.g., $11000100 \Rightarrow 11000000$) or mapped to the average of two equally likely solutions (e.g., $11010000 \Rightarrow 11100000$). Note that the following thermometer-to-binary conversion defines the sector $\Phi = 0$ at IQ-phases in the range $337.5° < \varphi < 0°$ (see also Fig. 4.46) because this definition requires the smallest number of logic gates. This is possible as the subsequent symbol decision ignores the absolute phase information and only evaluates relative phases.

4.3.4 Symbol Decision

Symbol decision is accomplished by differentiating the phase $\Phi(k)$, i.e., by calculating the phase difference from one symbol to the next, as shown in Fig. 4.53. This effectively measures the instantaneous frequency deviation $f_{dev}(k)$ with a quantization step of 62.5 kHz, where the sign of $f_{dev}(k)$ ideally carries the symbol information.

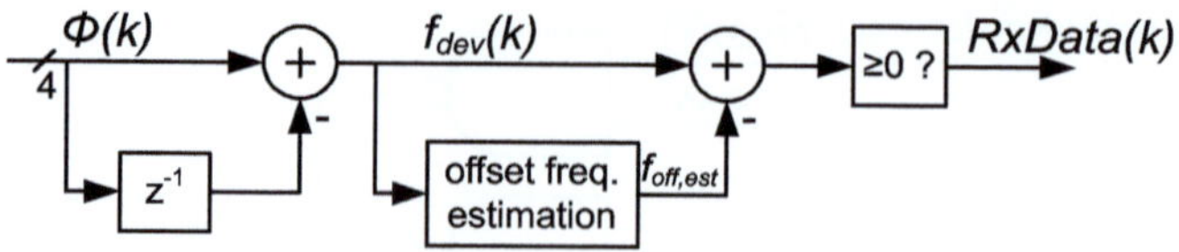

Fig. 4.53 Symbol decision implementation

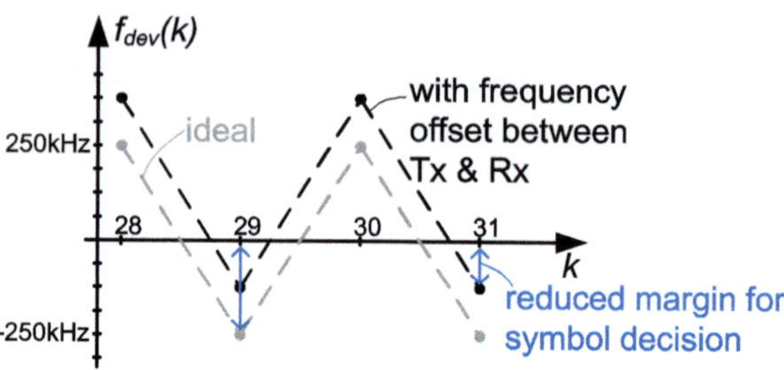

Fig. 4.54 Effect of carrier frequency offset on received frequency deviation $f_{\mathrm{dev}}(k)$ during the BLE preamble with its alternating bit sequence

In practice, the received frequency deviation $f_{\mathrm{dev}}(k)$ is biased by the carrier frequency offset between transmitter and receiver, which can be as large as $\pm150\,\mathrm{kHz}$ in a BLE system. This may reduce the margin for symbol decision from 4 to only 2 least significant bits (LSBs), as illustrated in Fig. 4.54. Hence, in order to make the demodulator robust against noise and intersymbol interference, the carrier frequency offset has to be compensated. This is done by subtracting from $f_{\mathrm{dev}}(k)$ an estimation of the carrier frequency offset $f_{\mathrm{off,est}}$. Such estimation is calculated by averaging the difference between the received and expected frequency deviations ($f_{\mathrm{dev}}(k)$ and $f_{\mathrm{dev,exp}}(k)$, respectively) during the last 4 bits of the 32-bit synchronization word after the BLE preamble. During this interval, the symbol clock should be already extracted with enough accuracy so as to obtain a close estimation of the frequency offset as,

$$f_{\mathrm{off,est}} = \frac{1}{4} \sum_{k=28}^{31} \left(f(k) - f_{\mathrm{dev,exp}}(k) \right). \tag{4.29}$$

where the value of $f_{\mathrm{dev,exp}}(k)$ is available given that the sync word is known a priori at the receiver. Afterward, $f_{\mathrm{off,est}}$ is stored and the averaging circuit disabled. No background offset calibration is required because the BLE frequency drift is limited to $\pm50\,\mathrm{kHz}$, which represents less than 1 LSB per symbol. Note that the clock extraction is currently not implemented in the transceiver, as already indicated in Fig. 4.40.

4.3.5 *Experimental Results*

The proposed low-power GFSK demodulator has been implemented in the same 130 nm CMOS process used before for the transceiver and frequency synthesizer.

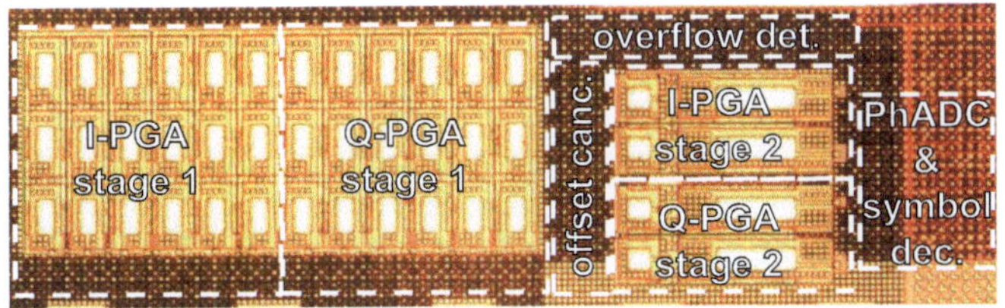

Fig. 4.55 Die photograph of the GFSK demodulator (zoom-in on the test chip 3 of Fig. 4.35). The occupied die size is $0.14\,\text{mm}^2$ (0.7 by 0.2 mm)

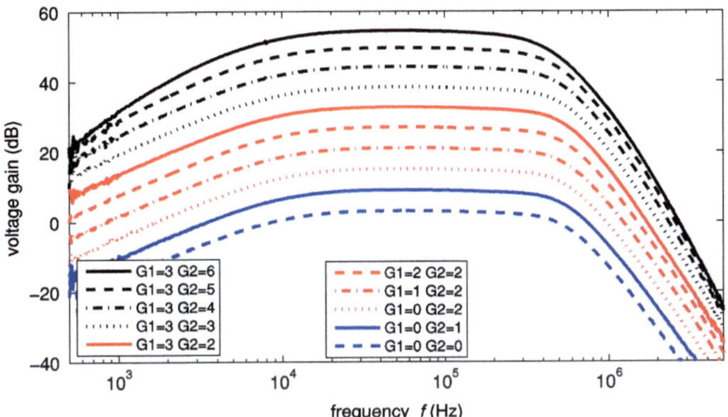

Fig. 4.56 Measured PGA voltage gain with selected gain settings G_1 and G_2 of the first and second PGA stage, respectively

Figure 4.55 shows the microphotograph of the die, where the MIM-capacitors of the two PGA-stages are the most distinctive elements. The area below these capacitors is used for the remaining passive elements, i.e., the large resistors of the PGA. Again, all measurements have been carried out at a supply voltage of 1 V, with the chip assembled in a QFN-36 package.

First, the performance of the PGA has been characterized using the test buffer at the voltage domain outputs of the PGA (see also Fig. 4.40). Figure 4.56 shows the measured voltage gain of the two-stage PGA with swept gain settings of the first and second stage, G_1 and G_2, respectively. In both cases, the 6 dB gain step and an almost constant transfer function can be observed. The 3 dB corner frequency variation is within $\pm10\,\%$ with respect to its nominal value 500 kHz, except for the maximum gain setting of the second stage $G_2 = 6$ where it is $-24\,\%$. Taking into account that this gain setting will only be applied for very weak input signals with a naturally low SNR, the reduced corner frequency is actually beneficial for demodulation because more noise will be filtered out.

The input referred noise spectrum of the PGA is dominated by flicker noise up to a frequency of approximately 120 kHz, shown in Fig. 4.57. In the small white noise

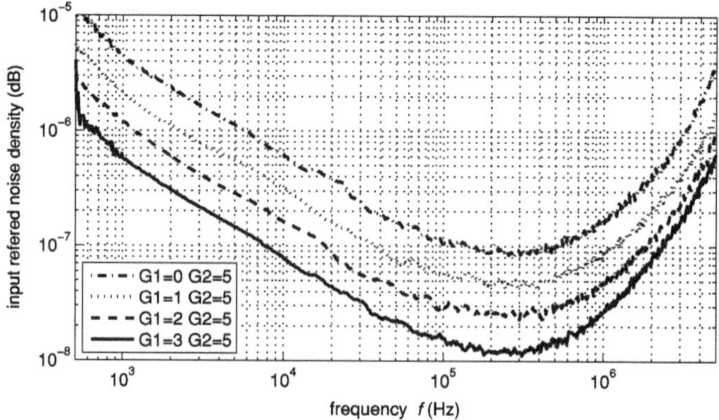

Fig. 4.57 Measured PGA input noise density

region around 300 kHz, the lowest input refereed noise density of $12.1\,\mathrm{nV}/\sqrt{\mathrm{Hz}}$ is achieved at the highest gain setting of the first PGA stage.

Next, the performance of the Ph-ADC is analyzed. As the generation of rotated phasors is accomplished in current-domain, there is no possibility to test the Ph-ADC alone, and hence measurements have been performed together with the PGA. First, the combination of PGA and Ph-ADC has been characterized with single input tones at 62.5 kHz, thus lying in the middle of the PGA pass-band. The phase shift of 90° between I- and Q-signal leads to a rotating phasor in counter-clockwise direction, giving rise to an increasing phase read-out as shown in Fig. 4.58. The analysis of this data obtains a maximum integral nonlinearity (INL) and a maximum differential nonlinearity (DNL) of 0.23 and 0.16 LSB, respectively.

Voltage- or current-domain ADCs are usually characterized by their effective number of bits (ENOB), calculated from measurements of the output signal-to-noise and distortion ratio (SNDR) for a single-tone input signal [8]. For a phase-domain ADC, an equivalent characterization is possible by applying a complex signal with a single-tone full-scale input phase $\varphi(t)$, i.e., an oscillating phase between $-\pi$ and π. However, such a signal has the disadvantage that not all the quantization levels are equally distributed and, hence, it does not adequately represent a noncoherent GFSK input signal. In order to use all phase quantization intervals equally, a complex input signal with constant magnitude and a two-tone phase $\varphi(t) = \pi \cos(2\pi f_1 t) + \pi \cos(2\pi f_2 t)$ is used for the ENOB characterization. Note that the corresponding spectra of the PGA input voltages $v_{I,\mathrm{in}}$ and $v_{Q,\mathrm{in}}$ contain more than two tones, as illustrated in Fig. 4.59a. These voltages are filtered and amplified by the PGA and phase-quantized by the Ph-ADC which ideally outputs a phase Φ containing only two tones again. Then, the ENOB can be calculated from the SNDR of the quantized phase Φ as follows

$$\mathrm{ENOB} = \log_2 \sqrt{\frac{\mathrm{SNDR}}{3}}. \qquad (4.30)$$

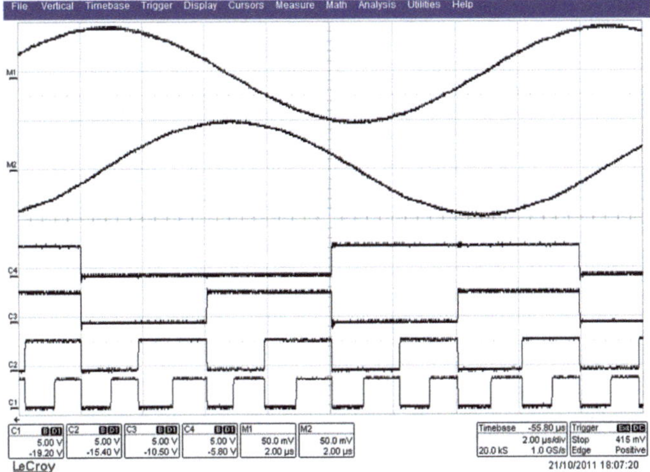

Fig. 4.58 Measured phase quantization of a rotating IQ-phasor (62.5 kHz). At the *top* the I- and Q-input voltages are shown and, below, the 4-bit phase output $\Phi(k)$ (sampled at 20 MHz)

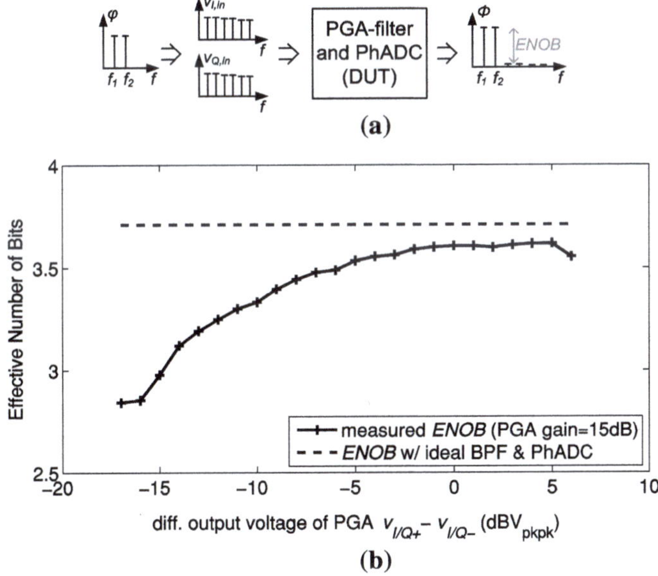

Fig. 4.59 Measurement of effective number of bits (ENOB) of the Ph-ADC. **a** Measurement concept with a two-tone input phase ($\varphi(t) = \pi \cos(2\pi f_1 t) + \pi \cos(2\pi f_2 t)$ with $f_1 = 25$ kHz and $f_2 = 123.1$ kHz). **b** Measured ENOB versus PGA output voltage

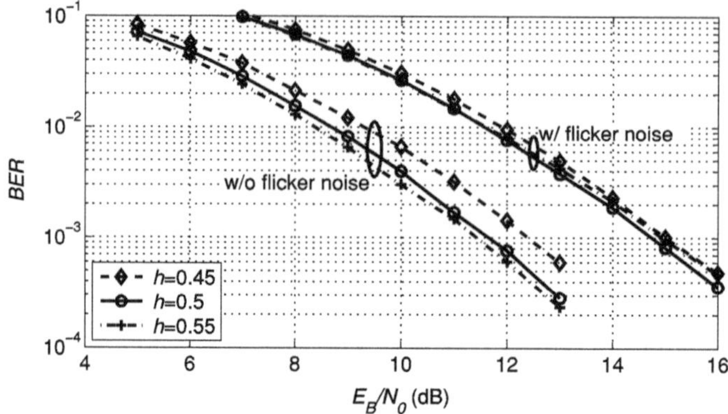

Fig. 4.60 Measured performance of the demodulator with and without flicker noise (corner frequency 150 kHz) for minimum, nominal, and maximum BLE modulation index h

The measured ENOB is shown in Fig. 4.59b versus the differential output voltage of the PGA. As expected, the ENOB increases with the amplitude because mismatch and noise become less dominant. The ENOB reaches a maximum of 3.61 bit and stays above 3.5 bit over a dynamic range of more than 10 dB. This wide dynamic range confirms that the PGA gain step size of 6 dB provides sufficient amplitude equalization. The difference between ENOB and the physical number of bits can be traced back to the bandpass characteristic of the PGA, and more specifically, to the nonconstant group delay caused by the HPF. Simulations of an ideal Ph-ADC, preceded by the implemented PGA (8–500 kHz passband), using a two-tone phase input already show a reduced ENOB. The two-tones $f_1 = 25$ kHz and $f_2 = 123.1$ kHz used in the measurements of Fig. 4.59b have been chosen because they lead to a high theoretically achievable ENOB of 3.71 and an almost uniform phase distribution due to the low greatest common divisor of 100 Hz.

In order to assess the performance of the Ph-ADC based demodulator, it has been tested with artificially generated noise. This noise is meant to represent the noise of the receiver frontend because this frontend usually dominates the system noise figure rather than the demodulator noise. To this end, the noisy baseband GFSK signal has been synthesized in MATLAB and then downloaded to the Agilent N5182A vector signal generator. This procedure allows not only for a precise control over the actual E_B/N_0 ratio but also for switching on and off the flicker noise. Figure 4.60 shows that assuming white noise only, an E_B/N_0-ratio of 11.7 dB is needed to obtain a BER of 0.1 % considering the nominal modulation index ($h = 0.5$). Adding the expected flicker noise with a corner frequency of 150 kHz, the required E_B/N_0 raises to 14.8 dB.

Figure 4.61 shows the performance of the demodulator assuming a carrier frequency offset f_{offset} between transmitter and receiver. With the compensation turned on, the BER performance is stable within the range of ± 150 kHz, as required by the BLE standard. Note that the BER performance without compensation is not

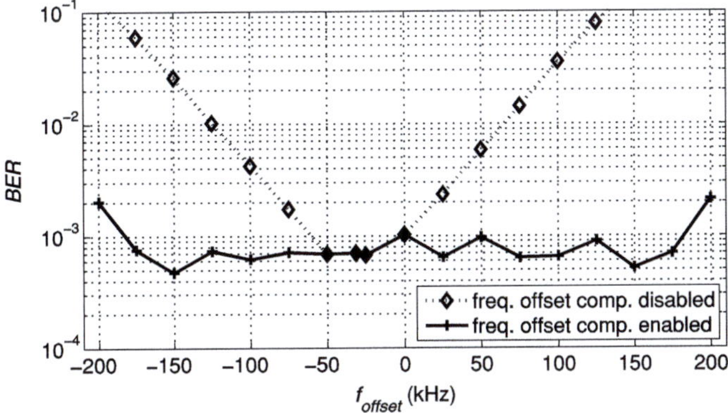

Fig. 4.61 Measured BER versus carrier frequency offset of the input signal with and without compensation ($E_B/N_0 = 15$ dB and considering flicker noise)

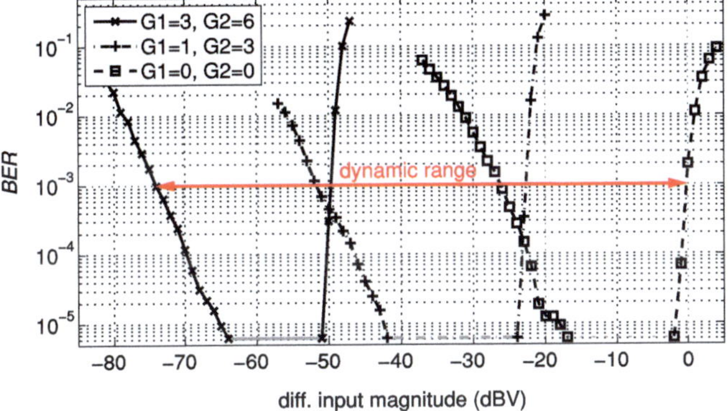

Fig. 4.62 Measured BER for selected gain settings of the two PGA stages illustrating the dynamic range of the demodulator of 74.0 dB

symmetric around a zero offset frequency but around a best case of -31.25 kHz which corresponds to a phase shift of half an LSB per symbol. This is caused by the inevitable nonsymmetric symbol decision that has to assign phase differences of $0°$ and $180°$ to either a decoded logical one or a zero, although both are equally likely.

Finally, in order to measure the sensitivity and the dynamic range of the complete demodulator including PGA, the demonstrator is tested with a noise-free GFSK signal at different input levels and gain settings, as shown in Fig. 4.62. A sensitivity of -74.1 dBV can be read from the performance with the maximum gain setting ($G_1 = 3$, $G_2 = 6$) considering again a BER of 0.1%. For the minimum gain setting ($G_1 = 0$, $G_2 = 0$), the prototype achieves correct demodulation up to a differential input signal level of -0.1 dBV giving a total dynamic range of 74 dB.

The measured performance of the demodulator is summarized in Table 4.6. Note that the demodulator complies with the requirements of the BLE standard with respect

Table 4.6 Measured performance summary of the zero-IF demodulator with phase-domain ADC

Parameter	Value
Technology	0.13 μm CMOS
Supply voltage	1.0 V
Current consumption	190 μA
PGAs	2×75 μA
Ph-ADC	25 μA
Symbol decision + auxiliary blocks	15 μA
Data rate, mod. index	1 Mb/s, $h = 0.5 \pm 10\%$
Carrier frequency offset	-170–170 kHz
Required E_b/N_0 (BER = 0.1 %)	
No flicker noise	11.7 dB
Flicker corner 150 kHz	14.8 dB
Sensitivity (BER = 0.1 %)	-74.1 dBV
Dynamic range	74.0 dB

Table 4.7 Comparison to recent GFSK demodulators

Parameter	[55]	[46]	[34]	This work[a]
Architecture	Low-IF	Low-IF	Low-IF	Zero-IF
CMOS technology (μm)	0.18	0.18	0.18	0.13
Data rate	1 Mb/s	250 kb/s	1 Mb/s	1 Mb/s
Modulation index	0.32	0.28	0.32	0.5
Power consumption	3.6 mW	630 μW	8.7 mW	190 μW
SNR @BER = 0.1 % (dB)	14.9	16.7	13.9	11.0[b]
Co-channel C/I	9.5 dB	–	9.1 dB	12.5 dB
Occupied chip area (mm^2)	0.26	0.11	0.26	0.14

[a] Includes PGA and channel filter

[b] Calculated in a bandwidth from 8 to 500 kHz and with flicker noise at a corner frequency of 150 kHz

to variations of modulation index and carrier frequency offsets. Due to the lack of published data from BLE demodulators, the comparison to the state-of-the-art in Table 4.7 relates this work to conventional Bluetooth demodulators. The proposed demodulator achieves with 190 μW the lowest power consumption and the lowest required signal-to-noise ratio (SNR), although the latter is also caused by the higher modulation index of BLE. The tolerated co-channel carrier-to-interference ratio (C/I) of 12.5 dB is about 3 dB worse than previously reported demodulators but still well below the BLE limit of 21 dB. The proposed demodulator achieves a low area occupation of only 0.14 mm^2 and offers more functionality because it includes the channel filtering PGA. This low chip area is possible due to the proposed scheme of phase generation in current domain which does not need large resistors.

4.3.6 Conclusions

A low-power GFSK demodulator tailored to the Bluetooth low energy standard has been presented in this section. It operates at zero-IF and so also allows for a simple and low power receiver architecture. The demodulator takes advantage of the constant envelope nature of the input signal and only quantizes the phase information by means of a phase-domain ADC with a resistor-less scheme of phase generation. Thanks to the phase generation in current domain, the prototype achieves both a low power consumption of $190\,\mu$W and a low area occupation of $0.14\,\text{mm}^2$, including the PGA and the channel filter. The presented demodulator requires an E_B/N_0 of $14.8\,$dB for a BER of $0.1\,\%$ considering a flicker noise corner of $150\,$kHz and so achieves a competitive performance compared to conventional low-IF demodulators. With a dynamic range of $74\,$dB and a carrier frequency offset tolerance of $\pm170\,$kHz it satisfies the requirements of BLE.

4.4 Overall Transceiver Performance

The measurements on the complete transceiver have been performed with chip 3 (Fig. 4.35). The same test board and antenna port configuration is employed, i.e., using a surface-mount balun with 2:1 impedance ratio [1] to provide a differential $100\,\Omega$ impedance for the chip.

The DC power consumption of the transceiver in TX-mode with all PA branches active is $P_{\text{DC,TX}} = 5.9\,$mW, which is dominated by the required DC power for the PA output stage, as shown in the power break down in Fig. 4.63a. The PA output stage consumes as much as $81\,\%$ of the total DC power and the PA drivers dissipate another $11\,\%$. The relative portion of the local oscillator (LO) is only $8\,\%$. With the RF output power of $P_{\text{out}} = 1.6\,$dBm, an overall power efficiency of $24.5\,\%$ is achieved. With the PA in power back-off operation, i.e., only one PA branch activated, the output power

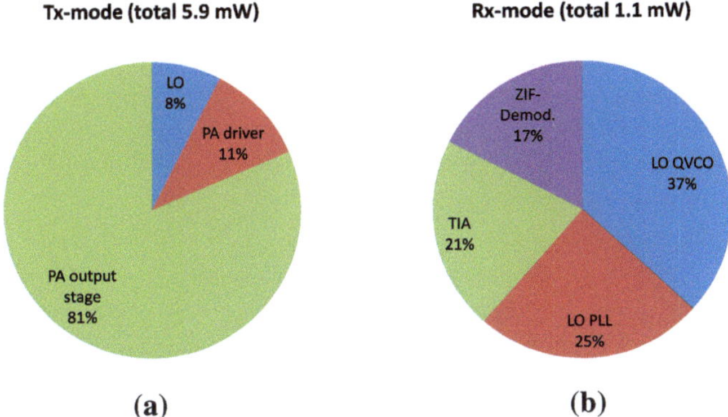

Fig. 4.63 Power break down of the transceiver in **a** TX mode and **b** RX-mode

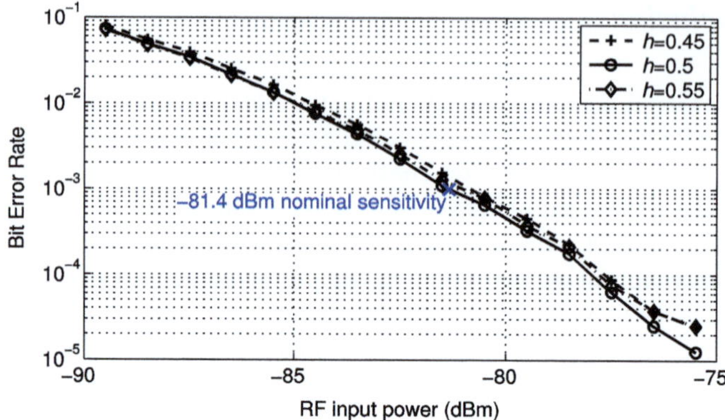

Fig. 4.64 Measured sensitivity of the receiver for minimum, nominal, and maximum BLE modulation index h, respectively

and DC consumption are reduced to -5 dBm and 2.9 mW, respectively, leading to an overall power efficiency of 10.8 %.

The total power dissipation of the transceiver in RX-mode is 1.1 mW. As shown in Fig. 4.63b, the largest portion of the power budget is taken by the QVCO, which consumes 400 mW. Note that this power consumption is ultimately limited by the LC-tank impedance level that can be implemented in the given technology. The complete frequency synthesizer, including the PLL, accounts for 62 % of the total power consumption and hence dominates the RX power budget. The remaining 38 % are distributed about equally to the first active amplification stage in the base-band, namely the TIAs at the output of the passive mixers, and the remaining base-band section, i.e., the zero-IF demodulator including the two-stage PGAs.

As shown in Fig. 4.64, the complete transceiver achieves a sensitivity of -81.4 dBm for a BER of 0.1 % and the nominal modulation index of $h = 0.5$. Within the tolerated variation of the BLE modulation index of ± 10 %, the sensitivity degrades by only 0.6 dB and, hence, exceeds the minimum BLE requirements by more than 10 dB in spite of the high noise figure of the receiver frontend. The receiver also fulfills the requirements on adjacent channel interference blocking, as shown in Table 4.8.

Table 4.9 compares this work to some recently published low-power transceivers operating in the 2.4 GHz ISM band. The presented transmitter achieves the highest efficiency with a power consumption comparable to that of OOK-transmitters. The receiver is compared based on the figure-of-merit

$$\text{FOM}_{\text{RX}} = P_{\text{DC,RX}} \cdot P_{\text{sens}}/R \tag{4.31}$$

where $P_{\text{DC,RX}}$ is the DC power consumption, P_{sens} is the sensitivity at 0.1 % BER, and R is the data rate [17]. A FOM_{RX} better than -200 dB are usually achieved by super-regenerative receivers, which are not tied to any specific WBAN standard. However, also the most recent BLE transceiver, presented by Wong et al. [141], demonstrates

Table 4.8 Measure RX performance

Parameter	Measured	BLE spec.
DC current (1.0 V supply)		
QVCO	400 µA	–
PLL	270 µA	–
TIAs	230 µA	–
Zero-IF demodulator	190 µA	–
Sensitivity	−81.4 dBm	< −70 dBm
Interference blocking co-channel	14.5 dB	<21 dB
@±1 MHz	1.1 dB	<15 dB
@±2 MHz	−17.6 dB	< −17 dB
@±3 MHz	−30.0 dB	< −27 dB
Tolerated carrier frequency offset	170 kHz	>150 kHz

Table 4.9 Comparison with recent low-power transceivers in the 2.4 GHz ISM band

Parameter	Raja'10 [104]	Contaldo'10 [37]	Vidojkovic'11 [133]	Wong'12 [141]	This work
CMOS Techn.	180 nm	180 nm + BAW	90 nm	130 nm	130 nm
Supply voltage	1.8 V	1.2 V/1.6 V	1 V/1.2 V	1.0 V	1.0 V
Standard	ZigBee	BLE	–	BLE	BLE
Modulation	OQPSK	GFSK ($h = 0.5$)	OOK	GFSK ($h = 0.5$)	GFSK ($h = 0.5$)
Data rate DR	250 kb/s	1 Mb/s	5 Mb/s	1 Mb/s	1 Mb/s
Channel spacing	5 MHz[a]	2 MHz	$> 4 \cdot DR$[b]	2 MHz	2 MHz
Transmitter					
Power cons. (mW)	18	47.3	3.7	4.6	2.9/5.9[c]
Output power (dBm)	0	5.4	−2	−10	−5/1.6[c]
Efficiency (%)	5.6	7.3	16.8	2.2	10.8/24.5[c]
Receiver					
Architecture	Low-IF	Heterodyne	Super-regen.	Sliding-IF	Zero-IF
Power cons.	22.3 mW	18.7 mW	534 µW	4.8 mW	1.1 mW
Sensitivity[d]	−94 dBm	−75 dBm@200 kb/s	−75 dBm	−94 dBm	−81.4 dBm
FOM_{RX}	−194.5 dB	−175.3 dB	−204.7 dB	−207.2 dB	−201.0 dB

[a] Multiple users per channel possible due to spectral spreading

[b] Selectivity reported as ±2 MHz at $DR = 500$ kb/s

[c] In power back-off/normal TX-mode

[d] For a bit error rate of 10^{-3}

a very good FOM_{RX} which is mainly due to its great sensitivity. On the other hand, in the subset of standard compliant solutions, the proposed passive frontend zero-IF receiver achieves clearly the lowest power consumption. In summary, the presented transceiver combines advantages of both groups, i.e., the low-power consumption of super-regenerative architectures with the spectral efficiency of BLE.

Finally, the presented transceiver has been successfully tested with a BLE master device nRF2739, which is part of a commercially available BLE development kit

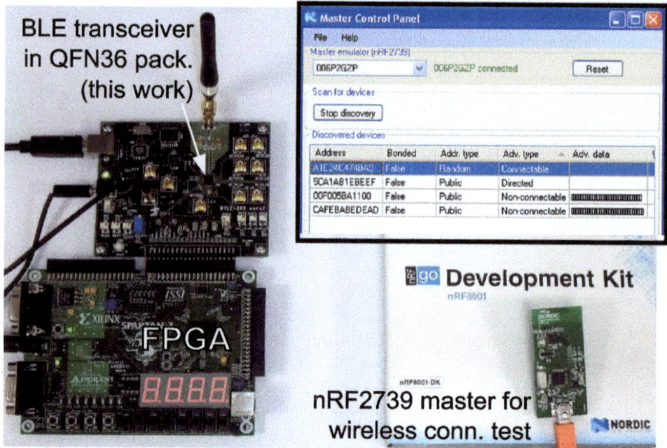

Fig. 4.65 Test of the wireless connection with a commercial BLE development kit. The inset shows the received BLE *Advertise* packets transmitted from the presented transceiver

[4], as shown in Fig. 4.65. The data stream for transmission as well as the control signals for opening and closing the PLL have been handled with an FPGA. In this test, the transceiver transmits different types of *Advertise* packets to the BLE master. As shown in the inset of Fig. 4.65, the packets are correctly conveyed to the BLE master. They include maximum length BLE frames of 376 bits (shown in the last two rows of the inset) thus, verifying the functionality of the direct QVCO modulation with the PLL opened.

4.5 Conclusions

A low power transceiver, fully compliant with the new Bluetooth low energy standard, has been presented. The main objective of the design has been to reduce its power consumption. Toward this goal, several power saving techniques have been implemented taking advantage of the relaxed specifications defined in the standard.

The transceiver, fabricated in a 130 nm CMOS technology, uses a zero-IF frontend receiver architecture. Only those active building blocks strictly needed to meet the receiver specifications have been considered. Accordingly, the receiver only comprises a step-up transformer to boost the internal RF impedance, passive mixers for direct down-conversion, and a low-power frequency synthesizer based on a double core quadrature VCO and a fast fractional-N PLL. Hence, the design does not include an LNA and the receiver frontend, excluding the frequency generation block, can be regarded as fully passive. The frequency synthesizer introduces a simple passive RC-network to cancel the magnetic coupling between the inductors of the QVCO. With this network, the IQ phase imbalance of the LO is kept below 1.5° with no power and negligible area cost. In the RX baseband, a 4-bit phase-domain ADC allows for a simple and power efficient GFSK demodulation. It employs a new linear combiner topology to generate the required phase rotations without resistors and so facilitates a compact implementation. Overall, the receiver achieves a power consumption as low as 1.1 mW, a sensitivity of -81.4 dBm at a data rate of 1 Mb/s, and a figure-of-merit similar to that obtained by recent super-regenerative receivers. Note, that whereas the spectral selectivity of these super-regenerative receivers is rather low, the proposed zero-IF receiver demodulates spectral efficient narrow-band GFSK ($h = 0.5$) signals.

The transmitter exploits the fact that BLE packets have a maximum length of 376 µs and employs a direct-modulation scheme in which the PLL is opened during transmission. Favored by the increased internal RF impedance provided by the step-up transformer, the transmitter includes an efficient class-E power amplifier which delivers an output power of 1.6 dBm to the differential antenna port. With a DC power consumption of 5.9 mW from a 1.0 V supply, the transmitter achieves a total power efficiency of 24.5 %.

Chapter 5
Co-integration of RF Energy Harvesting

One of the main challenges in the implementation of Wireless Body Area Networks (WBAN) is to make their element nodes energy autonomous so that they can be solely supplied by harvesting techniques [79]. Because of the varying operation conditions of transceivers (TRXs) for WBANs, energy harvesters have to be combined with the appropriate circuitry for charging storage elements, such as rechargeable batteries or supply capacitors [65, 73]. This chapter covers both aspects and presents an RF energy harvester, including supply management circuitry, which can be added to the previously proposed 2.4 GHz Bluetooth Low Energy transceiver.

Before discussing the possibilities to integrate RF energy harvesting, two important constraints have to be considered, i.e. the available RF input power levels and the characteristics of possible energy storage devices. Regarding the available RF energy, Visser et al. [134] have measured the ambient levels that can be expected in an office environment equipped with WLAN routers. According to their studies, the peak power density that can be expected is below 100 $\mu W/m^2$. Taking a look at the storage devices, solid-state thin-film rechargeable batteries provide a high energy density and low self-discharge currents [5, 6, 47, 56]. Nonetheless, such batteries usually require a minimum charging current on the order of a few tens of μAs [5, 6, 39]. Similarly, large supply capacitors with mF-capacitance, often referred to as super-capacitors, usually suffer from leakage currents on a similar order of magnitude [7, 74]. Hence, considering a supply voltage of 1V, charging a battery or supercapacitor requires a minimum DC power on the order of 25 μW [39] and, therefore, an incident RF power of about 100 μW (-10 dBm) is needed for RF energy harvesting if conversion losses are taken into account [65]. This excludes ambient RF energy as the sole source of energy, especially considering the small antenna sizes of wireless sensors [134, 136]. Nonetheless, in the close proximity of WLAN routers, which usually emit 20 dBm, sufficient power levels are present to charge the energy storage of a wireless sensor node. The advantage of such a charging scenario is that the sensor needs no connector for a wired charger.

Energy harvesting by RF-DC conversion may be incorporated into a wireless node by adding a dedicated antenna [53, 65, 93] or, more elegantly, by re-using the

J. Masuch and M. Delgado-Restituto, *Ultra Low Power Transceiver for Wireless Body Area Networks*, Analog Circuits and Signal Processing, DOI: 10.1007/978-3-319-00098-5_5,
© Springer International Publishing Switzerland 2013

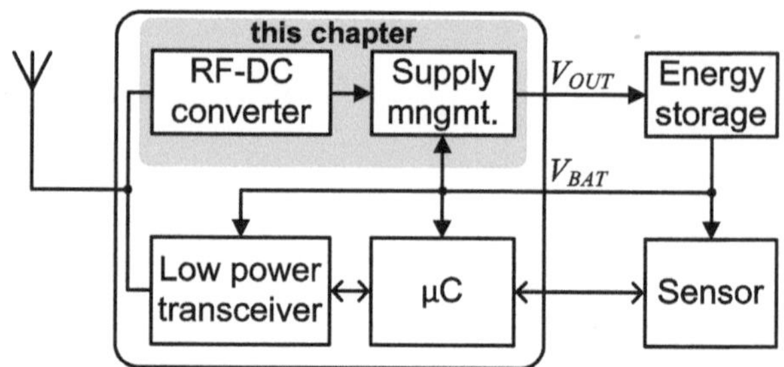

Fig. 5.1 Architecture of the wireless sensor node with RF energy harvesting

antenna of the TRX, as shown in Fig. 5.1. The latter facilitates system miniaturization but requires some kind of decoupling between the harvesting and communication operations. This can be accomplished by using two different carrier frequencies and a dual-band antenna [71]. However, this requires bulky LC matching networks to make the impedance looking into one functional unit high at the frequency the other functional unit is tuned to.

In the proposed approach, both the RF-DC conversion stage and the wireless link operate at the same carrier frequency and share the same single-band antenna. The output of the RF-DC converter drives a supply management circuit (SMC) to regulate the output voltage of a storage device. This circuit can be also configured as a constant current source, as it is usually required for driving Li-ion batteries [73]. The area consumption of the RF-DC conversion stage and the SMC is only 0.019 mm^2 in a 130 nm CMOS process. The harvester achieves an experimental power conversion efficiency (PCE) of 15.9 % at 0 dBm input power. This is better than the dual-band approach in [71], and also obtains a much smaller area occupation. Most importantly, the harvester has little impact on the transceiver performance; less than 0.5 dB degradation of output power and noise figure with respect to the previously implemented TRX frontend without the proposed harvester.

5.1 RF Switch for the Harvester

The frontend of the harvesting transceiver has to support three modes of operation, namely data transmission (TX), data reception (RX) and energy harvesting (RF-to-DC). The three modes are multiplexed such that only one is active at a time and the remaining two functional units are disconnected from the antenna by means of switches, as illustrated in Fig. 5.2. Considering the signal power ranges at which the respective functional unit operate, we see that mainly the TX has to be decoupled from the RF-to-DC block because the power ranges overlap. In other words, the switches avoid that the RF output power of the TX is converted back to DC.

Another important aspect regarding the switches is that different supply voltage conditions apply for them. Activating the TX- or RX-switch only makes sense when the supply voltage is sufficiently high to operate TX or RX, respectively. On the other

Fig. 5.2 RF switches

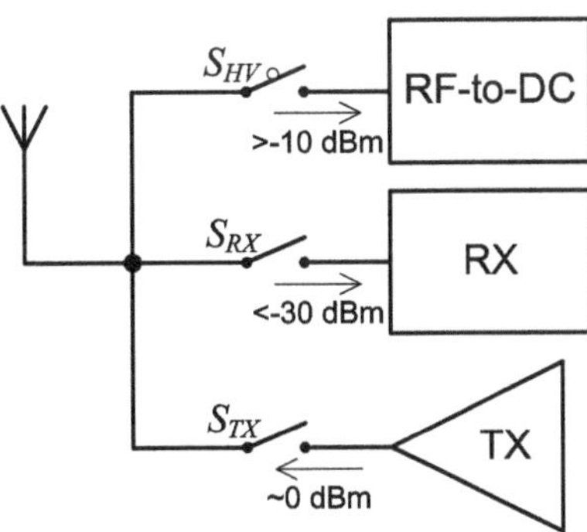

hand, the switch for the harvester should be turned on even if no supply voltage is available, i.e. if the energy storage of the wireless sensor is completely discharged.

In standard CMOS technologies the required functionality of the TX- and RX-switches can be easily implemented by means of enhancement-mode transistors [150]. The harvesting switch S_{HV}, however, requires an active-low behavior as exhibited by depletion-mode NMOS transistors which are able to form conducting channels even with negative gate-source voltages. As depletion-mode devices are not available in standard CMOS technologies [88], three different alternatives are discussed in the following.

5.1.1 LC-Resonator

One possibility to implement an active-low RF switch with an inherently active-high device, such as an enhancement-mode NMOS transistor, is by using resonators [137]. In the simple model shown in Fig. 5.3a the antenna is represented by a voltage source v_0 and a real impedance R_0. With the digital control voltage *HV_disable* being low, the transistor M_a is an open switch and the load R_{HV} is connected in series to the antenna through L_S and C_S, which ideally act as a short-circuit at resonance. When *HV_disable* is high, a parallel resonance circuit is formed by L_S and C_P, providing an ideally infinite input impedance Z_{in} at resonance.

Let us assume the harvester is active and the switch is characterized by its insertion loss factor ILF, defined as the ratio of the available power from the antenna to the power delivered to the load R_{HV}. Then, assuming matched source and load impedances ($R_{HV} = R_0$), the required quality factor of the series resonance circuit Q_S can be expressed as

$$Q_S = \frac{2\pi f_0 \cdot L_S}{2R_0 \cdot (\sqrt{\text{ILF}} - 1)} \tag{5.1}$$

where f_0 represents the resonance frequency. The de-activated switch may be assessed by quantifying how much the power transfer from the antenna to the RX,

Fig. 5.3 Alternatives for a single-ended active-low RF switch to connect the antenna (modeled as v_0 and R_0) to the harvester (R_{HV}): with LC-resonator (**a**), with $\lambda/4$-transmission line (**b**) and with start-up rectifier (**c**). To illustrate how the de-activated switch impairs the TX-performance the transmitter is shown in gray (v_{TX} and R_{TX} with an ideal switch S_{TX})

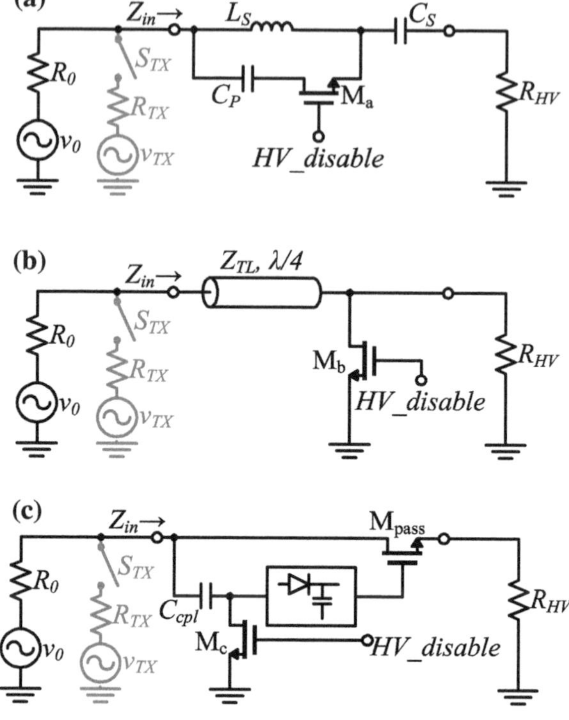

or equivalently from the TX to the antenna, is impaired due to leakage through the deactivated S_{HV} switch. As before, let us define the switch-induced degradation factor DGF as the ratio of available power from the source v_{TX} with its impedance R_{TX} to the power delivered to R_0. Then, assuming again matched conditions ($R_{\mathrm{TX}} = R_0$), the required quality factor of the parallel resonance circuit Q_P can be expressed as

$$Q_P = \frac{R_0}{4\pi f_0 \cdot L_S} \cdot \left(\frac{1}{\sqrt{\mathrm{DGF}} - 1} - 2 \right). \tag{5.2}$$

If we further assume equal quality factors of the serial and parallel resonators, they may be expressed as a function of the switch requirements.

$$Q_S = Q_P = \frac{1}{2} \cdot \sqrt{\frac{1}{\sqrt{\mathrm{ILF}} - 1} \cdot \left(\frac{1}{\sqrt{\mathrm{DGF}} - 1} - 2 \right)} \tag{5.3}$$

As an example, for an insertion loss of 1 dB (ILF = 1.26) and TX/RX-degradation of 0.5 dB (DGF = 1.12), the required quality factor of the resonators would be 5.5.

5.1.2 $\lambda/4$-Transmission Line

A second possibility to implement the active-low RF switch with an NMOS transistor is based on a $\lambda/4$-transmission line as shown in Fig. 5.3b. With *HV_disable* = 0 the load impedance is connected to the antenna through an ideally loss-less transmission line with a characteristic impedance Z_{TL} equal to load and source impedance ($Z_{TL} = R_{HV} = R_0$). When *HV_disable* is pulled high, the transistor M_b grounds the output of the transmission line which rotates this short-circuit one semi-circle across the Smith chart to an open-circuit. Hence, Z_{in} would ideally be infinite but in practice depends on the on-resistance $R_{DS,on}$ of the transistor [118].

$$Z_{in} = \frac{Z_{TL}^2}{R_{DS,on}||R_{Rect}} \approx \frac{Z_{TL}^2}{R_{DS,on}\,(M_b)} \tag{5.4}$$

5.1.3 Start-Up Rectifier

A third option for the active-low RF switch is to use an NMOS transistor as a pass device and generate the required positive gate-source voltage by means of a start-up rectifier, as shown in Fig. 5.3c. The start-up rectifier can have very small dimensions and does not require an outstanding efficiency, because it is loaded only capacitively by the gate of the pass transistor M_{pass}. However, in this concept the input power has to exceed a certain level given by the required turn-on voltage swing of the start-up rectifier. In order to de-activate the RF switch, the input of the rectifier is tied to ground by the transistor M_c and the antenna is mainly loaded capacitively by the coupling C_{cpl}.

5.1.4 Topology Selection

For selecting the most suitable topology, the transceiver frontend to be used for co-integrating the RF-DC conversion has to be taken into account. In this case, the previously proposed 2.4 GHz transceiver frontend is used, which is character- ized mainly by the high internal RF impedance of about 1 kΩ. The passive voltage gain due to impedance up-transformation is also expected to allow for an efficient RF-DC conversion since the rectifier performance usually improves for a higher input voltage swing [77].

The high RF impedance level excludes the *LC*-resonator approach in a fully- integrated solution as the required inductance would be approximately 90 nH accord- ing to (5.2) for DGF = 1.12, $Q_P = 5.5$ and $f_0 = 2.4$ GHz. Also the transmission line approach cannot be integrated into silicon at 2.4 GHz because the $\lambda/4$-transmission line would be about 16 mm long [118].

The solution with the start-up rectifier can be very area-efficient as no inductors or transmission lines are needed but it does not work for small signal levels because rectification relies on the non-linear large-signal characteristics of a diode. Anyhow, this is still a valid option for the given application if we keep in mind that we want to harvest RF energy only if the incoming power level is at least -10 dBm. This yields an input amplitude of at least 250 mV ($R_0 = 1$ kΩ) which is large enough to drive the start-up rectifier [65]. Therefore, the start-up rectifier solution is selected to co-integrate the harvester with the TRX frontend.

5.2 Architecture

The architecture of the low-power TRX frontend and the co-integrated RF energy harvester (highlighted on gray background) is shown in Fig. 5.4. The non-highlighted parts depict the relevant blocks of the transformer-based TRX frontend, which up-converts the antenna impedance to about 1 kΩ on chip. Note that the frontend already comprises the NMOS RF switches $MN_{1/2}$ and $MN_{3/4}$ to connect the power amplifier and passive mixer to the antenna, respectively.

In the modified frontend, the already existing RX-switch $MN_{3/4}$ is re-used as the switch for the RF energy harvester. This is possible because the mutual interaction between the passive mixer and the main rectifier, which performs the actual RF-to-DC conversion, can be kept low. A key aspect toward this goal is that the rectifier is an intrinsically nonlinear block whose parallel input resistance $R_{P,\mathrm{Rect}}$ strongly depends on the input power. Accordingly, the rectifier is designed such that it shows high input resistance at low input power level ($P_{\mathrm{RX}} < -20$ dBm) when the RX may be active and the mixer is matched to the transformer output. On the other hand, the rectifier exhibits a matched resistance ($R_{P,\mathrm{Rect}} \approx 1$ kΩ) when the RX is disabled and P_{RX} is typically around -5 dBm. Note that the reactive parts $C_{P,\mathrm{Rect}}$ and $C_{P,\mathrm{Mx}}$

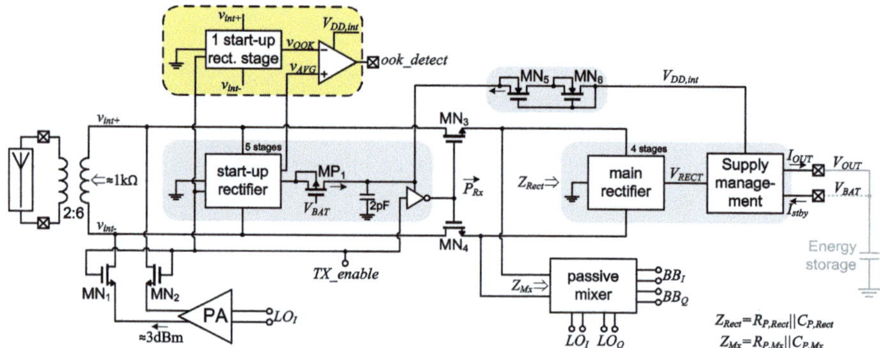

Fig. 5.4 Frontend of the low-power transceiver with the new parts for energy harvesting highlighted with gray background and the additional OOK detector in the *yellow dashed box*

Table 5.1 Decoupling strategy between the mixer and the rectifier

LO	$R_{P,\mathrm{Mx}}$	P_{RX}	$R_{P,\mathrm{Rect}}$	Dominant block
On	$\approx 1\,\mathrm{k\Omega}$	$< -20\,\mathrm{dBm}$	$>40\,\mathrm{k\Omega}$	Passive mixer
Off	$>20\,\mathrm{k\Omega}$	$\approx -5\,\mathrm{dBm}$	$\approx 1\,\mathrm{k\Omega}$	Main rectifier

are less significant because they do not dissipate power and can be tuned out by the transformer. Table 5.1 summarizes both operating conditions and shows that only one device is matched while the other presents high impedance. This approach also guarantees that the noise figure and sensitivity of the RX, both measured at low signal levels, are scarcely affected by the RF-DC converter.

In order to activate the RX-switch for energy harvesting without any external supply voltage, the start-up rectifier is directly connected to the internal transformer port $v_{\mathrm{int}\pm}$. Because of the small load that has to be charged by the start-up rectifier, it can be designed with much smaller dimensions than the main rectifier. However, in order to prevent the start-up rectifier from sinking the output power of the PA, its operation is controlled by an external signal, *TX_enable*, in such a way that when the PA is active, signal rectification is disabled and vice versa. Note that the start-up rectifier comprises 5 cascaded stages to guarantee that the RX-switch is on whenever there is significant RF input power to harvest, i.e. $P_{\mathrm{in}} > -10\,\mathrm{dBm}$.

By adding only 1 start-up rectifier stage more to the frontend, the harvesting frontend implements also an On-Off-Keying (OOK) detector. Hence, this additional feature comes with very little implementation cost but offers an additional RX-channel aside from the BLE receiver. This second channel can be used for example to send a wake-up signal to the sensor node. Such wake-up channels are frequently implemented in low power transceivers in order to save battery power and keep the main transceiver completely disabled until a wake-up signal is received [54, 63, 71, 100, 144]. Here, OOK-detection is achieved by comparing the output of the individual start-up rectifier stage v_{OOK} to the output of the first stage of the start-up rectifier v_{AVG}. As shown in Fig. 5.5, the only difference between these two rectifier stages is their loading, i.e. the former is loaded by the comparator only while the latter is loaded also by the remaining stages of the start-up rectifier and its load capacitance of 2 pF. Hence, v_{OOK} follows the OOK-envelope while v_{AVG} tracks the average of the envelope with a much larger time constant. Therefore, the OOK detection is simply performed by a comparator. Note that this comparator is designed with some intentional asymmetry in order to prevent the detection of false notches when the RF input signal is not modulated and hence $v_{\mathrm{OOK}} = v_{\mathrm{AVG}}$.

Finally, the supply management circuit transfers the DC output power of the main rectifier to the energy storage through the node V_{OUT}. An internal regulator limits this voltage to a maximum of 1.34 V to prevent oxide breakdown in the TRX. For a fast harvester start-up, the SMC is supplied with a small standby current I_{stby} through the node V_{BAT} to avoid the settling transients of the internal references, provided the energy storage is not empty. Nevertheless, with an empty energy storage, the harvester is still able to charge the load because I_{stby} is much smaller than I_{OUT}

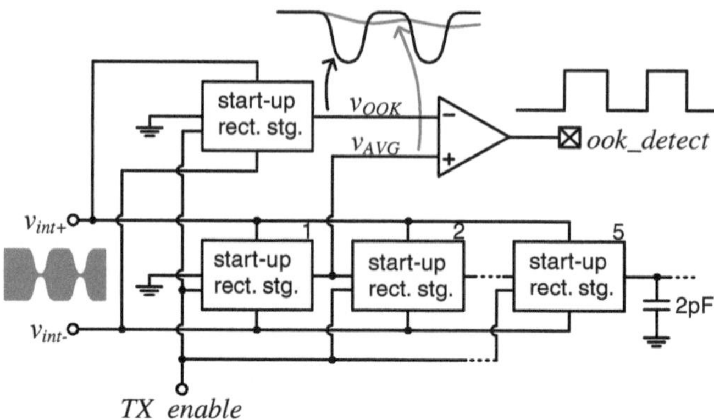

Fig. 5.5 Concept of the implemented OOK detection

under normal conditions. Also for speeding up the start-up, the SMC provides an internal supply $V_{DD,int}$ to pre-charge the RX-switch. Transistors $MN_{5/6}$ and MP_1 ensure the supply currents to flow in the desired directions only, indicated by arrows in Fig. 5.4. In this prototype, nodes V_{OUT} and V_{BAT}, which could be simply tied together to the load, use separate pins to allow for more flexibility and testability.

5.3 Circuit Design

5.3.1 Main Rectifier

The main rectifier is implemented as a cascade of 4 AC-coupled rectifier cells, each using the conventional 4-transistor configuration of Fig. 5.6 [49, 61, 77]. For input amplitudes below the threshold voltages, no current is flowing through the cell

Fig. 5.6 Schematic of one stage of the main rectifier (gate dimensions in μm)

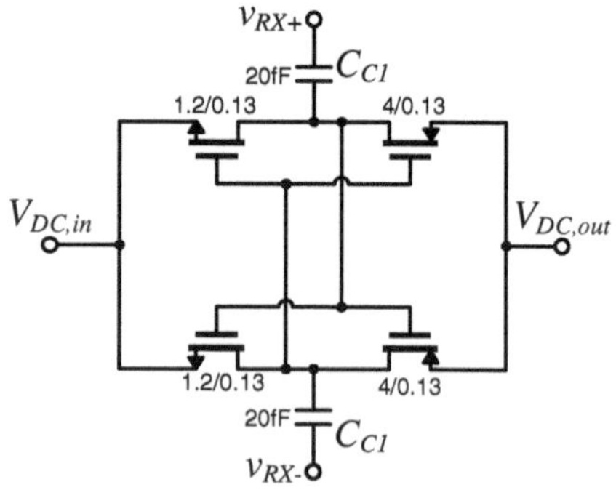

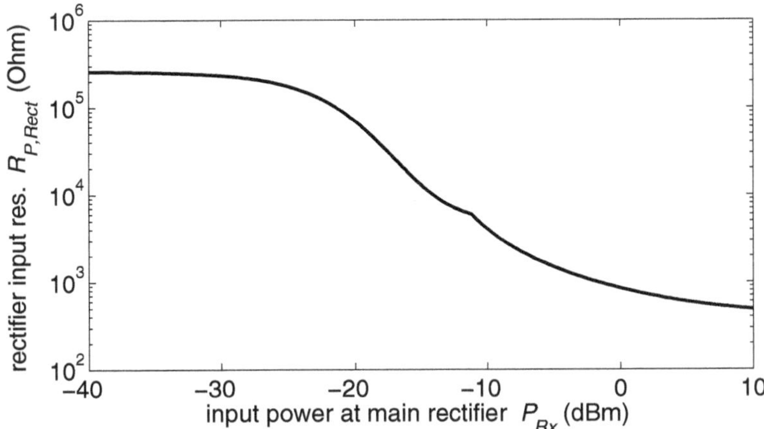

Fig. 5.7 Parallel input resistance $R_{P,\text{Rect}}$ of the main rectifier

transistors, thus leading to high $R_{P,\text{Rect}}$. As the input voltage increases, a periodic current starts flowing through the cell transistors and the input resistance of the structure diminishes. Figure 5.7 shows $R_{P,\text{Rect}}$ of the complete rectifier as a function of its input power, which closely follows the desired behavior expressed in Table 5.1. The main rectifier has been designed to match the 1 kΩ RF impedance at -5 dBm input power by choosing the transistor widths accordingly. To match the rectifier at lower input power levels, larger widths would be needed to reduce the input impedance.

5.3.2 Start-Up Rectifier

For the start-up rectifier completely different requirements apply. Since it is only capacitively loaded by the gates of the RX-switch, its efficiency is less important. It is much more critical to ensure that the start-up rectifier does not sink power from the PA when the TX is active. Accordingly, the 4-transistor rectifier cell is modified with an additional switch M_{off} to disable the start-up rectifier during transmission, as shown in Fig. 5.8. Pulling *disable* high, effectively shorts the two internal nodes and creates a virtual ground. In this mode, the RF input impedance of the modified rectifier cell is contributed by the series connection of the two coupling capacitors C_{C2} and the on-resistance of transistor M_{off}. In order to reduce power dissipation, M_{off} is made large enough so that the resistive part of the input impedance is small. On the other hand, when *disable* is low, M_{off} increases the parasitic load on the internal nodes of the rectifier leading to reduced efficiency. Accordingly, this concept is only suitable for the start-up rectifier but not for the main rectifier. The remaining devices are close to their respective minimum size in order to keep the capacitive loading of the RF nodes as low as possible.

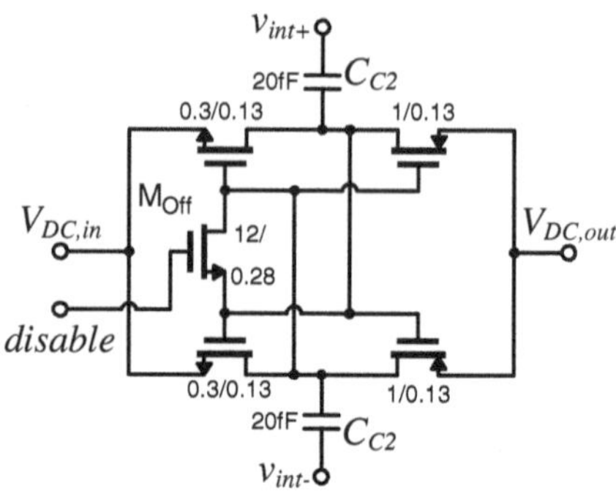

Fig. 5.8 Schematic of one start-up rectifier stage, which is also used for OOK-detection (gate dimensions in µm)

5.3.3 OOK-Comparator

The OOK-detector employs exactly the same rectifying stage as the start-up rectifier to track the envelope voltage of the incoming RF signal and compares this envelope voltage v_{OOK} to its average v_{AVG}, which is obtained from the start-up rectifier. To this end the comparator shown in Fig. 5.9 is implemented, which outputs a logical 1 whenever $v_{OOK} < v_{AVG} - V_{OS1}$, where V_{OS1} is the offset voltage of the comparator. In order to prevent the detection of false notches when $v_{OOK} = v_{AVG}$, the comparator employs an asymmetric PMOS input pair with different multiplicity-factors (m-factors) which guarantees a positive offset voltage V_{OS1} also taking into account device mismatch. The comparator is supplied with the internal supply voltage $V_{DD,int}$ and biased with a reference current of 30 nA, which are both obtained from the supply management circuit.

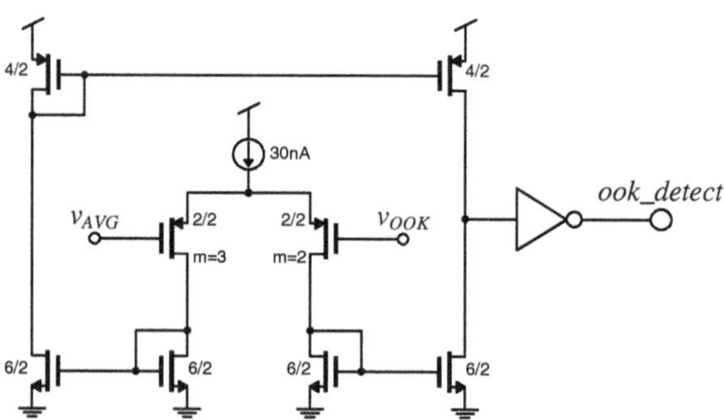

Fig. 5.9 Schematic of the OOK comparator (gate dimensions in µm)

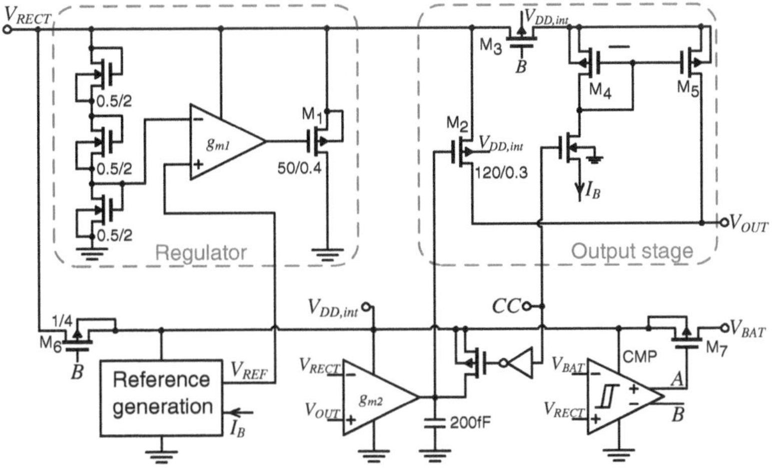

Fig. 5.10 Schematic of the supply management (transistor gate dimension in μm)

5.3.4 Supply Management Circuit

Figure 5.10 shows the schematic of the SMC. At its input, a regulator limits the rectified signal V_{Rect} to a DC voltage of 1.34 V, which is three times the value of the internal reference voltage V_{Ref}. The output stage of the SMC contains a PMOS pass transistor (M_2) which is on as soon as $V_{\text{Rect}} > V_{\text{OUT}} - V_{\text{OS2}}$, where V_{OS2} is the offset of the controlling amplifier g_{m2}. Similarly to the OOK comparator, this amplifier is designed such that its offset can only take negative values in order to prevent a reverse current through the pass device. When the control signal CC is high (constant current mode), the pass device is switched off and the current mirror ($M_{4/5}$) is activated, which outputs a constant current of about 70 μA. The switch M_3 prevents reverse current from V_{OUT} to V_{Rect} through the source-bulk diode of M_5.

The internal references are obtained from a CMOS-only reference generator without resistors [87] which only consumes 25 nA. It is supplied by the internal supply voltage $V_{\text{DD,int}}$ which is obtained either from V_{Rect} or V_{BAT}, depending on which is larger. The selection is accomplished by the comparator CMP and the switches $M_{6/7}$. The blocks connected to $V_{\text{DD,int}}$ also define the standby current I_{stby} drawn from the load, which add up to about 60 nA.

5.4 Experimental Results

The energy harvester, including the supply management circuit, has been integrated together with the 2.4 GHz low-power TRX frontend in the same 130 nm standard CMOS technology. Figure 5.11 shows the die microphotograph of this test chip 4.

Fig. 5.11 Microphotograph
of the assembled TRX with
energy harvester (test chip 4),
showing the start-up and the
main rectifier in the inset. The
total die size is 1.6 mm by
1.3 mm, equal to the TRX
w/o harvester (test chip 3,
Fig. 4.35)

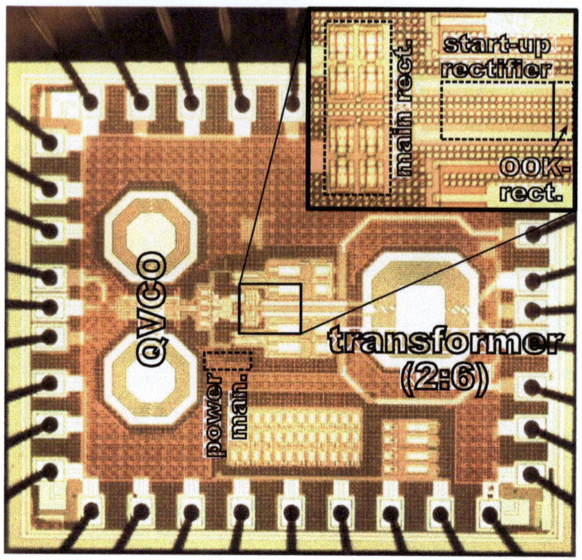

The area occupied by the energy harvester is 0.019 mm^2, which breaks down to
2100 µm^2, 4400 µm^2 and 12500 µm^2 for the start-up rectifier, the main rectifier and
the supply management circuit, respectively. For the measurements the chip has been
assembled again in a QFN-36 package and soldered onto a test board. Also, the same
antenna port configuration is employed, i.e. using a surface-mount balun with 2:1
impedance ratio [1] to provide a differential 100 Ω impedance for the chip.

Figure 5.12a illustrates the experimental performance of the unloaded supply
management circuit. As already mentioned, the output voltage becomes limited to
1.34 V when the regulator is active. Also note that with an external supply present
($V_{\mathrm{BAT}} = 1$ V), the RX-switch is always on and the harvester obtains a higher output
voltage at low input power levels ($P_{\mathrm{in}} \approx -15$ dBm). If $V_{\mathrm{BAT}} = 0$ V, the start-up
rectifier turns on the RX-switch for $P_{\mathrm{in}} > -11$ dBm.

Figure 5.12b shows the output current of the harvester in constant current mode
($CC = 1$) for different load voltages. Finally, Fig. 5.12c shows the measured PCE for
different load voltages without current regulation ($CC = 0$). The peak efficiencies
are obtained at about 0 dBm with a maximum of 15.9 % for a load voltage of
1.2 V. Therefore, the proposed energy harvester achieves better peak PCE than the
LC-decoupled harvester of Lerdsitsomboon et al. [71] but with a much smaller area.

Figure 5.13 shows the charging behavior of the harvester with a load capacitor
of 10 µF connected to both V_{OUT} and V_{BAT}. Although this load capacitance is too
small to serve as a real energy storage, it still allows to verify the dynamic performance of the harvester when pulsed RF energy is received as, for instance, provided
by a WLAN router. In Fig. 5.13 the results for two different RF pulse widths are
shown, representing the minimum and maximum WLAN packet length of 0.2 ms
and 5.6 ms, respectively. These measurements were also carried out at different input

Fig. 5.12 Measured performance of the energy harvester: **a** open circuit output voltage (10 MΩ load), **b** output current I_{OUT} in constant current mode ($CC = 1$), **c** power conversion efficiency (PCE) without current limitation ($CC = 0$) for different load conditions (V_{BAT} tied to V_{OUT})

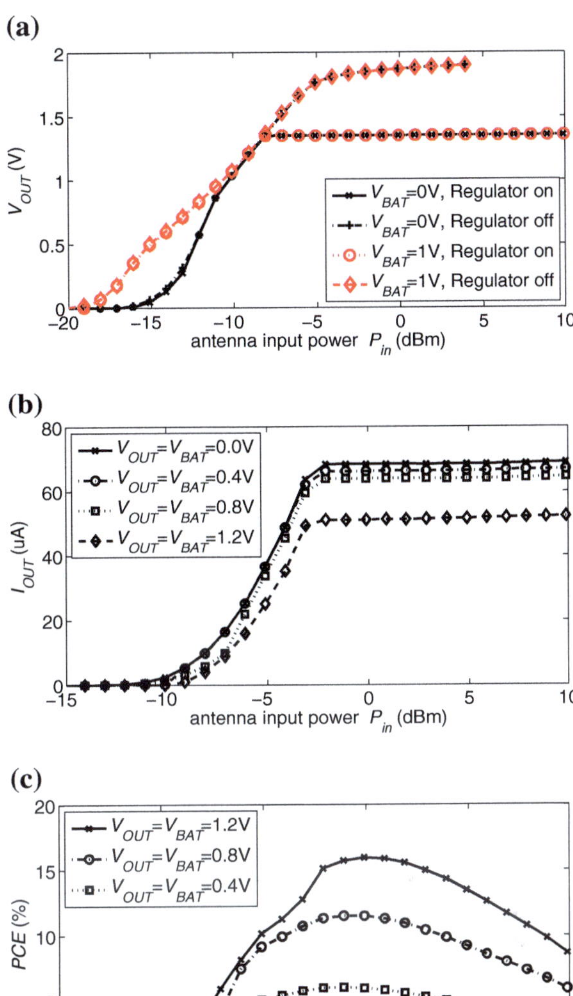

power levels. Figure 5.14 shows that for both pulse sequences, the harvester is able to progressively charge the 10 μF load until the regulator limits the voltage to 1.34 V, for input power levels of at least −9 dBm. Assuming isotropic antennas, this corresponds to an operation range of about 30 cm from a 20 dBm 2.4 GHz emitter in free space.

Figure 5.15 shows the operation of the OOK-detector without external supply voltage. In this example bursts of 10 notches with a notch width of 25 μs are transmitted

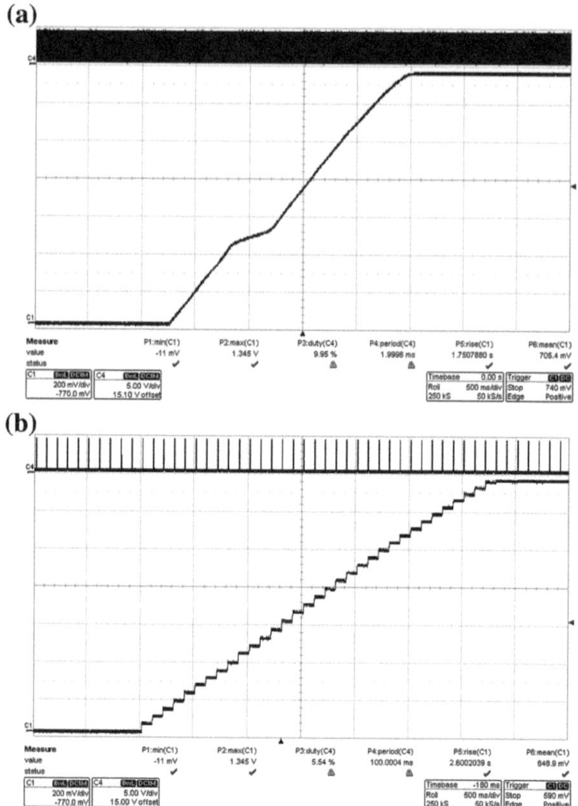

Fig. 5.13 Charging scenario with 10 μF load capacitance connected to V_{OUT} and V_{BAT} (*bottom curve*) with two different RF pulses sequences ($P_{in} = -2.1$ dBm, $f = 2.44$ GHz): **a** with a pulse width of 0.2 ms and a period of 2 ms, **b** with a pulse width of 5.6 ms and a period of 100 ms

and the RF input power is -12.6 dBm, which is the lowest power level that allows for correct notch detection. Note that due to the OOK-modulation, the rectified supply voltage $V_{OUT} = V_{BAT}$ drops to about 630 mV, which eventually limits the notch detection. This is because the comparator stops working correctly for lower supply voltages. The detector has been tested successfully for notch widths from 5 μs to 80 μs, which allows for data rates of up to 100 kb/s using pulse-interval encoding. With an external supply available, i.e. an energy storage which has been charged previously to 1.0 V, the minimum RF input level for correct OOK-detection drops to -19.6 dBm.

Figure 5.16 shows the input matching of the transceiver, both in energy harvesting mode and in RX-mode, measured through the balun with the HP8719D network analyzer. Note that in both modes the transceiver is well matched in the 2.4 GHz ISM band, i.e. the return loss is below -10 and -15 dB in harvest-mode and RX-mode, respectively.

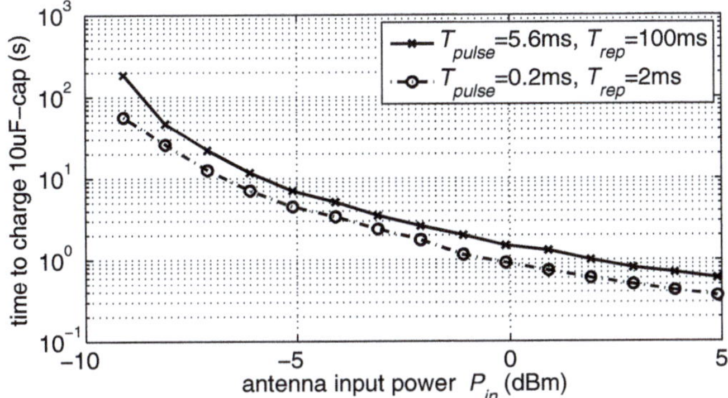

Fig. 5.14 Measured 10–90 % rise time for charging a 10 μF storage capacitor to the final voltage of 1.34 V

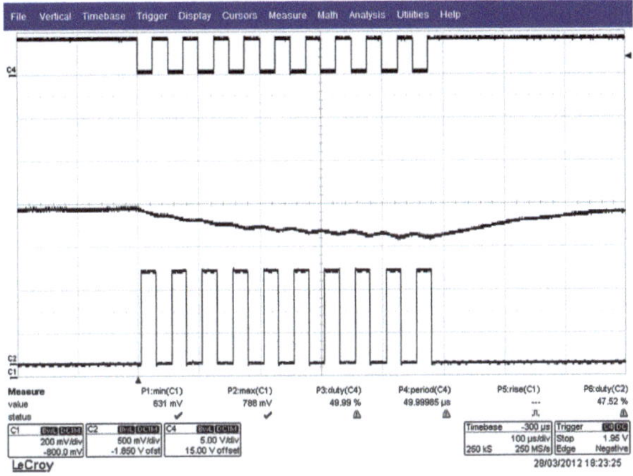

Fig. 5.15 Operation of the OOK-demodulator without external supply and an RF input power of −12.6 dBm: at the top the modulating for generating the OOK-signal with a signal generator is shown, below the self-generated supply $V_{\text{OUT}} = V_{\text{BAT}}$ connected (using the common load capacitance of 10 μF) and at the bottom the detected OOK-notches *ook_detect*

Finally, the performances of the two TRX frontends (test chip 3 and 4), which only differ on the presence or not of the proposed energy harvester, are compared. Figure 5.17 shows the measured noise figure (NF) and output power P_{out} for both chips. Due to process variations between MPW runs, the TRX with harvester achieves its best performance at a slightly higher frequency. This also explains the slightly higher peak output power for the TRX with harvester. Regarding the NF, the TRX is only degraded by about 0.2 dB with the harvester co-integrated. These results are in good agreement with post-layout simulation results that predicted a degradation of 0.4 dB and 0.2 dB for NF and P_{out}, respectively.

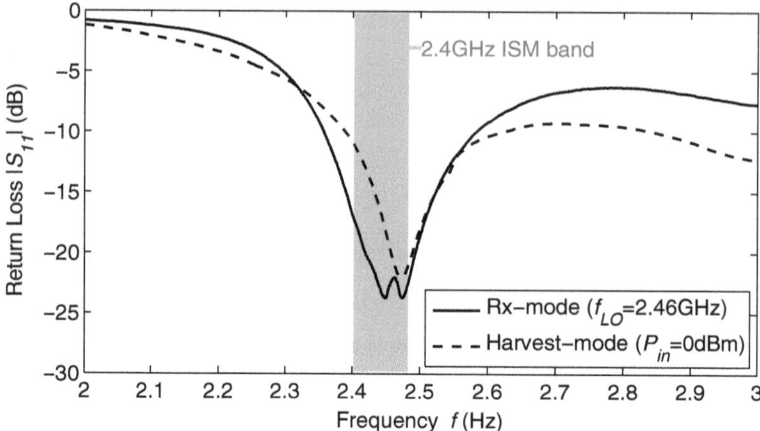

Fig. 5.16 Return loss $|S_{11}|$ in RX-mode

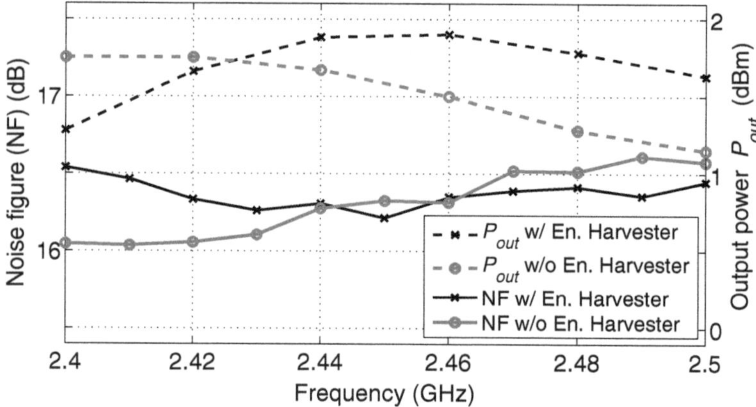

Fig. 5.17 Measured noise figure (NF) and output power P_{out} of the implemented TRX frontend with and without energy harvester

5.5 Conclusions

In this chapter the low-power transceiver frontend has been supplemented with an energy harvester that can be easily co-integrated, where both the TRX and the harvester operate at the same frequency of 2.4 GHz. Decoupling between the TX and the harvester is achieved by an RF-switch that is turned on without the need for an external supply, i.e. using a start-up rectifier. The decoupling between the RX and the harvester exploits the different input impedances of these blocks. The harvester achieves a measured peak PCE of 15.9 % and degrades the TX- and RX-performance by less than 0.5 dB. Since the required area for the harvester is only 0.019 mm^2, it can be added to the TRX frontend at almost no cost.

Table 5.2 Comparison with recent RF energy harvester co-integrated with TRX

Parameter	Ishizaki'10 [58]	Lerdsitsomboon'11 [71]	This work
Implemented system	RX+harvester	TRX+harvester	TRX+harvester
CMOS technology (nm)	90	130	130
Frequency (GHz)	2.4	2.4 & 5.8	2.4
Decoupling strategy	actively tuned *RC*-network	*LC*-network	passively activated NMOS switch
TRX-degradation	not reported	0.3–0.5 dB	0.4 dB (RX) 0.2 dB (TX)
Peak PCE @ power level (dBm)	15.8% @ −2	14% @ 0	15.9% @ 0
Harvests w/o ext. supply	✗	✓	✓
Wake-up channel Sensitivity	✗	✓ −13 dBm (passive mode)	✓ −12.6 dBm (passive mode) −19.6 dBm (battery-assisted)
Area (mm^2)	≈0.08	0.24 [a]	0.019

[a] An additional bond wire inductor is needed

Table 5.2 summarizes the performance of the energy harvester and compares it to recently published RF energy harvesters that share a common antenna with a transceiver or receiver. However, note that the harvester of Ishizaki et al. [58] requires a supply voltage for harvesting in order to tune the input matching network to the RF signal to be harvested. The proposed harvester with the start-up rectifier achieves a similar efficiency as the two other solutions. Also in terms of the degradation of the transceiver and the sensitivity of the wake-up channel similar results are obtained compared to Lerdsitsomboon et al. [71]. However, the main advantage of the proposed solution is that it occupies a much lower die area and does not need any bond wire inductors.

Chapter 6
Conclusions

In this work, an ultra-low power transceiver architecture has been presented, which is able to fulfill the Bluetooth low energy standard. Especially with respect to the RX power consumption, the proposed transceiver advances the state of the art, achieving a DC power dissipation of 1.1 mW, as shown in Fig. 6.1. It is the lowest power consumption reported for a narrow-band transceiver in the 2.4 GHz ISM band, which fulfills one of the typical WBAN standards. With a sensitivity of -81.4 dBm and the resulting RX figure-of-merit of -201.0 dB, the receiver achieves a competitive performance also taking into account proprietary wide-band and pulsed ultra wide-band implementations. Note that in all figures of this chapter narrow-band, wide-band and IR-UWB transceivers are distinguished by blue, black, and red markers, respectively. The low-power consumption is mainly due to the passive RX frontend architecture in conjunction with the maximized impedance level at the few remaining RF nodes. Moreover, the low complexity phase-domain demodulator leads to a power efficient base-band section.

Concerning the performance in TX-mode, the power consumption of 5.9 mW in normal mode and 2.9 mW in back-off mode is among the lowest reported so far for narrow-band transmitters, as shown in Fig. 6.2. However, the total transmitter efficiency is significantly higher with respect to the state of the art, meaning that a higher portion of the dissipated power is converted into transmitted signal power. For the most part, the good efficiency is again due to the maximized internal RF impedance level. Also the direct QVCO-modulation while opening the PLL plays a decisive role in the low-power consumption.

Overall, the complete transceiver sets a new reference point concerning the power consumption of narrow-band WBAN transceivers. It achieves a wireless link budget of 83.0 dB while consuming together only 7 mW in the transmitter and receiver, as shown in Fig. 6.3. It is, therefore, highly suitable for applications with limited energy resources such as biomedical wireless sensors.

It has also been shown that the selected front-end architecture can be easily equipped with an RF energy harvester using only one single-band antenna. Measurements have confirmed that this additional feature degrades the transceiver

J. Masuch and M. Delgado-Restituto, *Ultra Low Power Transceiver for Wireless Body Area Networks*, Analog Circuits and Signal Processing, DOI: 10.1007/978-3-319-00098-5_6,
© Springer International Publishing Switzerland 2013

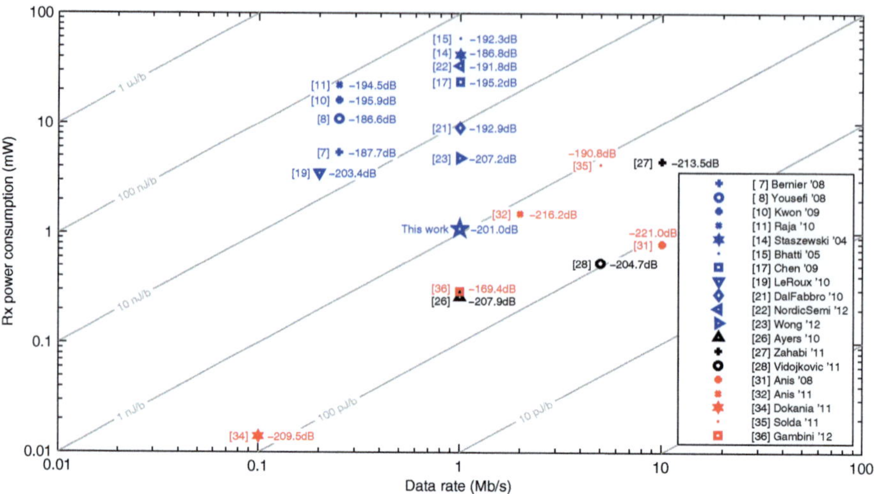

Fig. 6.1 Comparison to the receive performance of recent low-power transceivers. The dB-values in the figure denote the figure-of-merit FOM_{RX} of the respective receiver, as defined in (4.31)

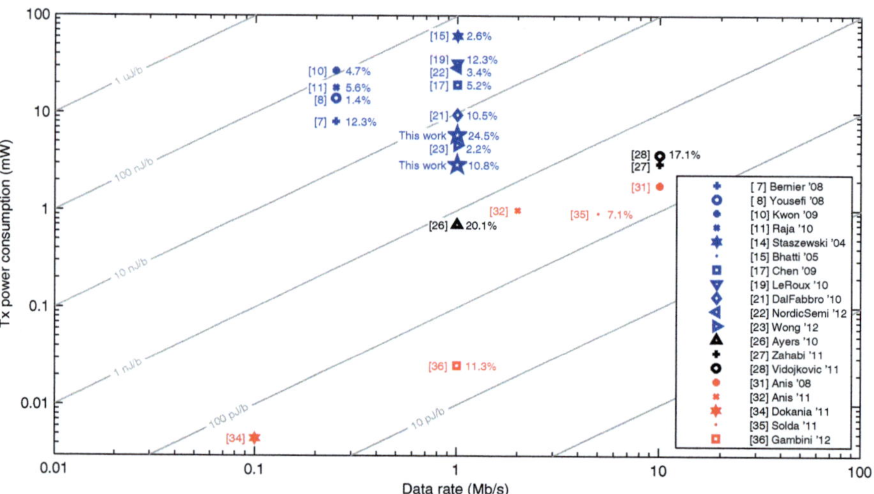

Fig. 6.2 Comparison to the transmit performance of recent low-power transceivers. The percentages in the figure denote the total power efficiency of the respective transmitter ($P_{out}/P_{DC,TX}$).

performance by less than 0.5 dB and can therefore be implemented at almost no cost. The harvester achieves a decent peak efficiency of 15.9 % and is able to progressively charge up an energy storage device for pulsed input signals emitted by an active WLAN router, for an expected distance of up to approximately 30 cm.

Apart from the system level, this book presents the following contributions at the circuit-level:

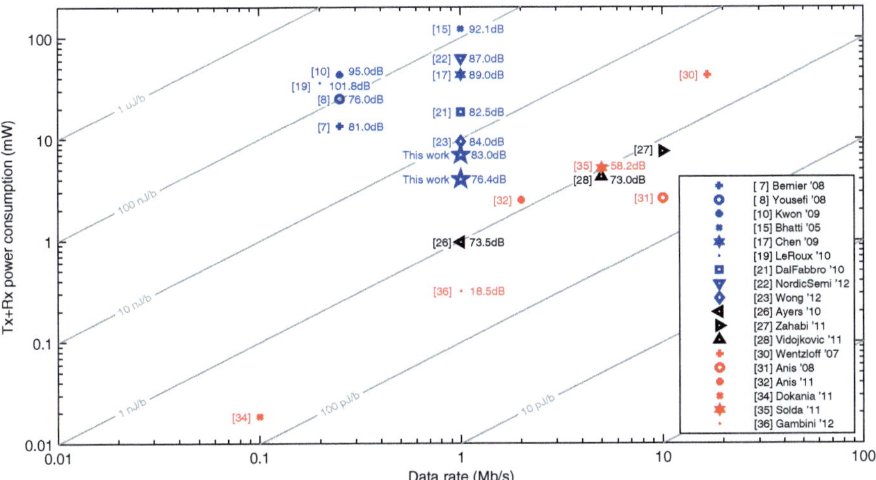

Fig. 6.3 Comparison to recent low-power transceivers considering the sum of TX and RX power consumption (y-axis). The dB-values in the figure denote the link budget of the respective transceiver ($P_{out} - P_{sens}$)

1. A passive RC-network to cancel the magnetic coupling between the two inductors of the QVCO with negligible area demand and a coupling factor suppression of at least 12 dB considering the worst case technological corners.
2. A simple (G)FSK modulation cell comprised two differential PMOS transistor pairs to modulate the QVCO-tank capacitance. Although the cell requires calibration with respect to process and mismatch variations, it has been verified to be sufficiently stable for the BLE signaling within the industrial temperature range ($-40\,°C \ldots + 85\,°C$).
3. A phase-domain ADC with a new topology to generate the phase-shifted signals using a current-domain linear combiner. As opposed to conventional linear combiners based on resistors, the new topology allows for a both area- and power-efficient implementation.
4. An RF switch, which can be passively turned on, is presented to decouple the RF energy harvester from the transmitter. In harvesting mode a start-up rectifier enables the switch using the incoming RF energy and in TX-mode the start-up rectifier is short circuited in order to prevent loading of the transmitter.

6.1 Future Work

Although the presented transceiver contains all fundamental blocks and functions to modulate and demodulate BLE signaling, a few auxiliary blocks still have to be added to obtain a stand-alone transceiver. For example, the reference clock should be generated internally by a crystal oscillator and the RX-symbol clock should be extracted from the BLE preamble in future implementations. However, none of these blocks are expected to increase the power consumption significantly. Also the digital functions that are currently implemented externally in the FPGA, such as the calibration of the GFSK modulation index and the control of the PGA gain, have to be integrated on chip.

Moreover, taking into account that both the TX and RX power consumption is ultimately limited by the integratable Q-factor of the inductive elements (the TX by the transformer and the RX by the QVCO inductors), possibilities to improve these elements may be considered. For example, a silicon on insulator (SOI) technology may be an option as it reduces the substrate losses of those elements.

Regarding the wireless sensor node as a whole, research on other sources for energy harvesting would be interesting, given the power and distance limitations of RF energy harvesting. To obtain complete energy autonomy, schemes exploiting temperature differences or kinetic energy may be a promising option.

To provide connectivity to subcutaneously implanted sensors, another interesting option is the addition of a wireless interface suitable for communication through tissue with a high-water-content. Together with an interface based on magnetic coupling or using the body as a communication channel (body-channel communication, BCC), the presented transceiver could serves as a relay device between implanted sensors and a smartphone or other external units.

Appendix
Radio Specifications Imposed by the BLE Standard

The Bluetooth low energy (BLE) standard is an extension of the conventional Bluetooth standard which was published with the version 4.0 of the latter in late 2009 [3]. The BLE extension is intented for small devices with limited energy resources allowing for short range communication with other BLE devices or with so called dual-mode device supporting both conventional Bluetooth and BLE. A typical dual-mode device would be a smartphone which already support conventional Bluetooth and only requires minor modifications to implement BLE as well, for example through a firmware update. On the other hand, the energy-constrainted small devices would be single-mode devices which only support BLE in order to talke full advantage of the relaxed radio specifications.

With the BLE standard, the energy demand of single-mode transceivers is expected to reduce by as much as 90 % compared to conventional Bluetooth basic rate. This is made possible by simplification of the protocol in the link layer in order to reduce the time needed for connection set-up. This will especially reduce the average power consumption of devices with a low duty cycle, which is typically the case for wireless sensors. But also the in the physical layer, the radio specifications have been relaxed to facilitate power savings. For example, the channel spacing for Bluetooth low energy is larger and the requirements to block interferers from adjacent channels is reduced. Table A.1 summarizes the radio specification of the BLE standard in contrast to Bluetooth basic rate.

It is instructive to translate the radio specifications defined by the BLE standard into requirements for the spectral purity of the frequency synthesizer. The spectral purity of an oscillator is measured as the phase noise which is defined as the ratio of noise power to carrier power in a small bandwidth (usually 1 Hz) at a given offset frequency Δf from the carrier. It is measured as a single-sideband value, i.e. measured at the upper sideband of the carrier with positive offset frequencies.

To calculate the tolerable phase noise level, a certain demodulation performance, i.e. a required signal-to-noise ratio (SNR) to achieve the targeted bit-error-rate (BER), has to be assumed. A GFSK demodulators for a modulation index of $h = 0.5$ requires an SNR of about 11 dB, according to theoretic analysis [68]. To allow for

J. Masuch and M. Delgado-Restituto, *Ultra Low Power Transceiver for Wireless Body Area Networks*, Analog Circuits and Signal Processing, DOI: 10.1007/978-3-319-00098-5,

Table A.1 Bluetooth radio specifications for the basic rate and for the BLE extension

Parameter	Bluetooth basic rate	Bluetooth low energy
RF channels	$f = 2402 + k$ MHz, $k = 0, 1, ..., 78$	$f = 2402 + 2k$ MHz, $k = 0, 1, ..., 39$
Carrier frequency tolerance	± 75 kHz	± 150 kHz
Maximum drift rate	400 Hz/µs	400 Hz/µs
Maximum drift per package	± 40 kHz	± 50 kHz
Longest package length	3.1 ms	376 µs
Modulation type	GFSK	GFSK
Modulation index h	$0.32 \pm 9\%$	$0.5 \pm 10\%$
Data rate	1 Mb/s	1 Mb/s
Spurious emissions[a]		
Adjacent channel (2 MHz)	-20 dBm	-20 dBm
Adjacent channel (≥ 3 MHz)	-40 dBcm	-30 dBm
Required sensitivity[b]	-70 dBm	-70 dBm
Carrier-to-interference ratio (CIR)[b]		
Co-channel	11 dB	21 dB
Channel (1 MHz)	0 dB	15 dB
Channel (2 MHz)	-30 dB	-17 dB
Channel (≥ 3 MHz)	-40 dB	-27 dB
Exeption for image channel	-9 dB	-9 dB

[a] Measured in 1 MHz bandwidth
[b] For a BER = 0.1

sufficient implementation margin, let us assume that an SNR of 15 dB is required for the following calculations. Then, a constraint can be formulated that considers the integrated phase noise in a given bandwidth BW around the carrier.

$$\frac{1}{\text{SNR}} > 2 \int_0^{\text{BW}/2} \mathscr{L}(\Delta f) \cdot d\Delta f \tag{A.1}$$

Assuming a constant phase noise within the channel bandwidth the required in-channel average phase noise $\mathscr{L}_{\text{in-chan}}$ for a $BW = 1$ MHz channel simplifies to

$$\mathscr{L}_{\text{in-chan}} < \frac{1}{\text{SNR} \cdot \text{BW}} = -75 \text{ dBc/Hz}. \tag{A.2}$$

In a similar manner also the phase noise requirements at larger offset from the carrier can be calculated by considering the adjacent channel CIR at offset channel frequencies f_{ch}.

$$\int_{f_{\text{ch}}-0.5\text{MHz}}^{f_{\text{ch}}+0.5\text{MHz}} \mathscr{L}(\Delta f) \cdot d\Delta f < \frac{\text{CIR}(f_{\text{ch}})}{\text{SNR}} \tag{A.3}$$

The requirement can be simplified again by assuming a constant average phase noise in the adjacent channel. However, at larger offset frequencies the phase noise usually decreases with a certain slope. Therefore, the constraint for the adjacent channel phase noise will be formulated at the lower end of the bandwidth representing the worst case. The average tolerable phase noise in the adjacent channels with 2 MHz and 3 MHz offset is consequently denoted as $\mathscr{L}_{1.5\text{MHz}}$ and $\mathscr{L}_{2.5\text{MHz}}$, respectively.

$$\mathscr{L}_{1.5\text{MHz}} < \frac{\text{CIR}(2\text{MHz})}{\text{SNR} \cdot \text{BW}} = -92\,\text{dBc/Hz} \tag{A.4}$$

$$\mathscr{L}_{2.5\text{MHz}} < \frac{\text{CIR}(3\text{MHz})}{\text{SNR} \cdot \text{BW}} = -102\,\text{dBc/Hz} \tag{A.5}$$

The requirement based on the adjacent channel at 1 MHz offset is intentionally not mentioned here because the large allowed carrier to interference ratio of 15 dB is more related to the selectivity of the baseband filters of the receiver than to a practical phase noise requirement of the synthesizer.

The tolerable spur level according to the spurious emission specification is $-20\,\text{dBc}$ or $-30\,\text{dBc}$ at 2 MHz or 3 MHz offset, respectively, assuming a 0 dBm transmitter. However, a more stringent requirement for the spur level arises again from the receiver specification. The tolerable spur level normalized to the carrier ($P_{\text{spur}}/P_{\text{carrier}}$) can be derived by considering how much an interferer at a channel offset f_{ch} has to be suppressed to fall below a certain signal to noise ratio.

$$P_{\text{spur},f_{ch}}/P_{\text{carrier}} < \frac{\text{CIR}_{f_{ch}}}{\text{SNR}} \tag{A.6}$$

$$P_{\text{spur},2\text{MHz}}/P_{\text{carrier}} < -32\,\text{dBc} \tag{A.7}$$

$$P_{\text{spur},3\text{MHz}}/P_{\text{carrier}} < -42\,\text{dBc} \tag{A.8}$$

References

1. Chip-Balun WE-BAL 748421245 specification for release (2004)
2. AVR2006: Design and characterization of the Radio Controller Board's 2.4GHz PCB Antenna (2007)
3. Bluetooth Specification, Volume 6 (Low Energy Controller) (2009)
4. Development Kit for nRF8001 (nRF8001-DK) (2010)
5. EnerChip CBC050 Datasheet Rev B, Rechargeable Solid State Energy Storage (2010)
6. Printed Thin Film Battery (2010)
7. GW109 / GW209 Supercapacitor Datasheet Rev 1.0 (2011)
8. IEEE Standard for Terminology and Test Methods for Analog-to-Digital Converters. IEEE Std 1241–2010 (Revision of IEEE Std 1241–2000) pp. 1–139 (2011)
9. nRF8001 Single-chip Bluetooth low energy solution, Product Specification 1.0 (2012)
10. Abouei, J., Brown, J.D., Plataniotis, K.N., Pasupathy, S.: Energy Efficiency and Reliability in Wireless Biomedical Implant Systems. IEEE Transactions on Information Technology in Biomedicine 15(3), 456–466 (2011). (Kostas)
11. Andreani, P., Bonfanti, A., Romano, L., Samori, C.: Analysis and design of a 1.8-GHz CMOS LC quadrature VCO. IEEE Journal of Solid-State Circuits 37(12), 1737–1747 (2002)
12. Andreani, P., Mattisson, S.: On the use of MOS varactors in RF VCOs. IEEE Journal of Solid-State Circuits 35(6), 905–910 (2000)
13. Andreani, P., Xiaoyan, W.: On the phase-noise and phase-error performances of multiphase LC CMOS VCOs. IEEE Journal of Solid-State Circuits 39(11), 1883–1893 (2004)
14. Anis, M., Ortmanns, M., Wehn, N.: A 2.5mW 2Mb/s fully integrated impulse-FM-UWB transceiver in 0.18μm CMOS. In: IEEE MTT-S Int. Microwave Symp. Dig., pp. 1–3 (2011)
15. Anis, M., Tielert, R., Wehn, N.: A 10Mb/s 2.6mW 6-to-10GHz UWB impulse transceiver. In: IEEE Int. Conf. on Ultra-Wideband, vol. 1, pp. 129–132 (2008)
16. Annamalai Arasu, M., Kok Fong Ong, H., Yeung Bun, C., Wooi Gan, Y.: A 2.4-GHz CMOS RF front-end for wireless sensor network applications. In: IEEE Radio Frequency Integr. Circuits Symp. Dig., pp. 455–458 (2006)
17. Ayers, J., Mayaram, K., Fiez, T.S.: An Ultralow-Power Receiver for Wireless Sensor Networks. IEEE Journal of Solid-State Circuits 45(9), 1759–1769 (2010)
18. Ayers, J., Panitantum, N., Mayaram, K., Fiez, T.S.: A 2.4GHz wireless transceiver with 0.95nJ/b link energy for multi-hop battery-freewireless sensor networks. In: Symp. on VLSI circuits, pp. 29–30 (2010)
19. Bae, J., Yan, L., Yoo, H.J.: A Low Energy Injection-Locked FSK Transceiver With Frequency-to-Amplitude Conversion for Body Sensor Applications. IEEE Journal of Solid-State Circuits 46(4), 928–937 (2011)

J. Masuch and M. Delgado-Restituto, *Ultra Low Power Transceiver for Wireless Body Area Networks*, Analog Circuits and Signal Processing, DOI: 10.1007/978-3-319-00098-5, © Springer International Publishing Switzerland 2013

20. Balankutty, A., Shih-An, Y., Yiping, F., Kinget, P.R.: A 0.6-V Zero-IF/Low-IF Receiver With Integrated Fractional-N Synthesizer for 2.4-GHz ISM-Band Applications. IEEE Journal of Solid-State Circuits **45**(3), 538–553 (2010)

21. Banerjee, B., Enz, C.C., Le Roux, E.: A 290μA, 3.2MHz 4-bit phase ADC for constant envelope, ultra-low power radio. In: NORCHIP Conf., pp. 1–4 (2010)

22. Banerjee, B., Enz, C.C., Le Roux, E.: Detailed analysis of a phase ADC. In: Proc. IEEE Int. Symp. on Circuits and Systems, pp. 4273–4276 (2010)

23. Belmas, F., Hameau, F., Fournier, J.: A Low Power Inductorless LNA With Double G_m Enhancement in 130 nm CMOS. IEEE Journal of Solid-State Circuits **47**(5), 1094–1103 (2012)

24. Bernier, C., Hameau, F., Billiot, G., de Foucauld, E., Robinet, S., Durupt, J., Dehmas, F., Mercier, E., Vincent, P., Ouvry, L., Lattard, D., Gary, M., Bour, C., Prouvee, J., Dumas, S.: An ultra low power 130nm CMOS direct conversion transceiver for IEEE802.15.4. In: IEEE Radio Frequency Integr. Circuits Symp. Dig., pp. 273–276 (2008)

25. Betancourt-Zamora, R.J., Verma, S., Lee, T.H.: 1-GHz and 2.8-GHz CMOS injection-locked ring oscillator prescalers. In: Symp. on VLSI circuits, pp. 47–50 (2001)

26. Bhatti, I., Roufoogaran, R., Castaneda, J.: A fully integrated transformer-based front-end architecture for wireless transceivers. In: IEEE Int. Solid-State Circuits Conference Dig. Tech. Papers, pp. 106–587 Vol. 1 (2005)

27. Bo, X., Chunyu, X., Wenjun, S., Valero-Lopez, A.Y., Sanchez-Sinencio, E.: A GFSK demodulator for low-IF Bluetooth receiver. IEEE Journal of Solid-State Circuits **38**(8), 1397–1400 (2003)

28. Borremans, J., Ryckaert, J., Desset, C., Kuijk, M., Wambacq, P., Craninckx, J.: A Low-Complexity, Low-Phase-Noise, Low-Voltage Phase-Aligned Ring Oscillator in 90 nm Digital CMOS. IEEE Journal of Solid-State Circuits **44**(7), 1942–1949 (2009)

29. Byun, S., Park, C.H., Song, Y., Wang, S., Conroy, C.S.G., Kim, B.: A low-power CMOS Bluetooth RF transceiver with a digital offset canceling DLL-based GFSK demodulator. IEEE Journal of Solid-State Circuits **38**(10), 1609–1618 (2003)

30. Calvo, B., Celma, S., Martinez, P.A., Sanz, M.T.: 1.8 V-100 MHz CMOS programmable gain amplifier. In: Proc. IEEE Int. Symp. on Circuits and Systems, p. 4 pp. (2006)

31. Carrara, F., Presti, C.D., Pappalardo, F., Palmisano, G.: A 2.4-GHz 24-dBm SOI CMOS Power Amplifier With Fully Integrated Reconfigurable Output Matching Network. IEEE Transactions on Microwave Theory and Techniques **57**(9), 2122–2130 (2009)

32. Chang, G., Jansson, L., Wang, K., Grilo, J., Montemayor, R., Hull, C., Lane, M., Estrada, A.X., Anderson, M., Galton, I., Kishore, S.V.: A direct-conversion single-chip radio-modem for Bluetooth. In: IEEE Int. Solid-State Circuits Conference Dig. Tech. Papers, vol. 1, pp. 88–448 vol. 1 (2002)

33. Chee, Y.H., Niknejad, A.M., Rabaey, J.M.: An Ultra-Low-Power Injection Locked Transmitter for Wireless Sensor Networks. IEEE Journal of Solid-State Circuits **41**(8), 1740–1748 (2006)

34. Chen, C.P., Yang, M.J., Huang, H.H., Chiang, T.Y., Chen, J.L., Chen, M.C., Wen, K.A.: A Low-Power 2.4-GHz CMOS GFSK Transceiver With a Digital Demodulator Using Time-to-Digital Conversion. IEEE Transactions on Circuits and Systems–Part I: Regular Papers 56(12), 2738–2748 (2009)

35. Cho, S., Chadrakasan, A.P.: A 6.5-GHz energy-efficient BFSK modulator for wireless sensor applications. IEEE Journal of Solid-State Circuits **39**(5), 731–739 (2004)

36. Cojocaru, C., Pamir, T., Balteanu, F., Namdar, A., Payer, D., Gheorghe, I., Lipan, T., Sheikh, K., Pingot, J., Paananen, H., Littow, M., Cloutier, M., MacRobbie, E.: A 43mW Bluetooth transceiver with -91dBm sensitivity. In: IEEE Int. Solid-State Circuits Conference Dig. Tech. Papers, pp. 90–480 vol. 1 (2003)

37. Contaldo, M., Banerjee, B., Ruffieux, D., Chabloz, J., Le Roux, E., Enz, C.C.: A 2.4-GHz BAW-Based Transceiver for Wireless Body Area Networks. IEEE Transactions on Biomedical Circuits and Systems **4**(6), 391–399 (2010)

38. Cook, B.W., Berny, A., Molnar, A., Lanzisera, S., Pister, K.S.J.: Low-Power 2.4-GHz Transceiver With Passive RX Front-End and 400-mV Supply. IEEE Journal of Solid-State Circuits **41**(12), 2757–2766 (2006)

39. Costinett, D., Falkenstein, E., Zane, R., Popovic, Z.: RF-powered variable duty cycle wireless sensor. In: European Microwave Conf., pp. 41–44 (2010)
40. Craninckx, J., Steyaert, M.S.J.: A 1.75-GHz/3-V dual-modulus divide-by-128/129 prescaler in 0.7-μm CMOS. IEEE Journal of Solid-State Circuits **31**(7), 890–897 (1996)
41. Crepaldi, M., Chen, L., Fernandes, J.R., Kinget, P.R.: An Ultra-Wideband Impulse-Radio Transceiver Chipset Using Synchronized-OOK Modulation. IEEE Journal of Solid-State Circuits **46**(10), 2284–2299 (2011)
42. Dal Fabbro, P.A., Pittorino, T., Kuratli, C., Kvacek, R., Kucera, M., Giroud, F., Tanner, S., Chastellain, F., Casagrande, A., Descombes, A., Peiris, V., Farine, P.A., Kayal, M.: A 0.8V 2.4GHz 1Mb/s GFSK RF transceiver with on-chip DC-DC converter in a standard 0.18μm CMOS technology. In: Proc. European Solid-State Circuits Conference, pp. 458–461 (2010)
43. De Muer, B., Steyaert, M.: CMOS Fractional-N Synthesizers. Kluwer Academic Publishers, Analog circuits and signal processing (2003)
44. Dehaese, N., Bourdel, S., Barthelemy, H., Bachelet, Y., Bas, G.: FSK zero-crossing demodulator for 802.15.4 low-cost receivers. In: IEEE Int. Conf. on Electronics, Circuits and Systems, pp. 1–4 (2005).
45. Dokania, R.K., Wang, X.Y., Tallur, S.G., Apsel, A.B.: A Low Power Impulse Radio Design for Body-Area-Networks. IEEE Transactions on Circuits and Systems–Part I: Regular Papers **58**(7), 1458–1469 (2011)
46. Dong, H., Yuanjin, Z.: An ultra low power GFSK demodulator for wireless body area network. In: Proc. European Solid-State Circuits Conference, pp. 434–437 (2008)
47. Dudney, N.J.: Solid-state thin-film rechargeable batteries. Materials Science and Engineering B **116**(3), 245–249 (2005)
48. Enz, C., Scolari, N., Yodprasit, U.: Ultra low-power radio design for wireless sensor networks. In: Proc. IEEE Int. Workshop Radio-Frequency Integration Technology: Integrated Circuits for Wideband Communication and Wireless Sensor, Networks, pp. 1–17 (2005)
49. Facen, A., Boni, A.: Power Supply Generation in CMOS Passive UHF RFID Tags. In: Conf. on Ph.D. Research in Microelectronics and, Electronics, pp. 33–36 (2006)
50. Gambini, S., Crossley, J., Alon, E., Rabaey, J.H.: A Fully Integrated, 290 pJ/bit UWB Dual-Mode Transceiver for cm-Range Wireless Interconnects. IEEE Journal of Solid-State Circuits **47**(3), 586–598 (2012)
51. Gupta, M., Bang-Sup, S.: A 1.8-GHz Spur-Cancelled Fractional-N Frequency Synthesizer With LMS-Based DAC Gain Calibration. IEEE Journal of Solid-State Circuits **41**(12), 2842–2851 (2006)
52. Haddad, F.: Zai, x, d, L., Sangiovanni, A., Scali, R.: Low cost and low power open-loop frequency synthesiser. Electronics Letters **46**(14), 982–984 (2010)
53. Haizheng, G., Sobot, R.: RF power harvesting analog front-end circuit for implants. In: Canadian Conf. on Electrical and, Computer Engineering, pp. 907–910 (2009)
54. Hambeck, C., Mahlknecht, S., Herndl, T.: A 2.4μW Wake-up Receiver for wireless sensor nodes with -71dBm sensitivity. In: Proc. IEEE Int. Symp. on Circuits and Systems, pp. 534–537 (2011)
55. Hong-Sing, K., Ming-Jen, Y., Tai-Cheng, L.: A Delay-Line-Based GFSK Demodulator for Low-IF Receivers. In: IEEE Int. Solid-State Circuits Conference Dig. Tech. Papers, pp. 88–589 (2007)
56. Hu, L., Wu, H., La Mantia, F., Yang, Y., Cui, Y.: Thin, Flexible Secondary Li-Ion Paper Batteries. ACS Nano **4**(10), 5843–5848 (2010). doi:10.1021/nn1018158
57. Hu, W.Y., Lin, J.W., Tien, K.C., Hsieh, Y.H., Chen, C.L., Tso, H.T., Shih, Y.S., Hu, S.C., Chen, S.J.: A 0.18-μm CMOS RF Transceiver With Self-Detection and Calibration Functions for Bluetooth V2.1 + EDR Applications. IEEE Transactions on Microwave Theory and Techniques **58**(5), 1367–1374 (2010)
58. Ishizaki, H., Mizuno, M.: A 0.2 mm^2, 27 Mbps 3 mW ADC/FFT-Less FDM BAN Receiver With Energy Exploitation Capability. IEEE Journal of Solid-State Circuits **45**(4), 921–927 (2010)

59. Khan, T., Raahemifar, K.: A low power current reused quadrature VCO for biomedical applications. In: Proc. IEEE Int. Symp. on Circuits and Systems, pp. 669–672 (2009)
60. Kluge, W., Poegel, F., Roller, H., Lange, M., Ferchland, T., Dathe, L., Eggert, D.: A Fully Integrated 2.4-GHz IEEE 802.15.4-Compliant Transceiver for ZigBee™ Applications. IEEE Journal of Solid-State Circuits 41(12), 2767–2775 (2006)
61. Kotani, K., Sasaki, A., Ito, T.: High-Efficiency Differential-Drive CMOS Rectifier for UHF RFIDs. IEEE Journal of Solid-State Circuits 44(11), 3011–3018 (2009)
62. Kwon, I., Song, S.S., Eo, Y.: A fully integrated 2.4-GHz CMOS RF transceiver for low-rate wireless sensor networks. Microwave and Optical Technology Letters 51(6), 1481–1487 (2009)
63. Kwon, Y.I., Byun, J., Jo, H.J., Lee, H.Y.: An ultra low power CMOS transceiver having a short range wake-up for IR-WPAN applications. In: European Microwave Integr. Circuits Conf., pp. 262–265 (2009)
64. Kwon, Y.I., Park, S.G., Park, T.J., Cho, K.S., Lee, H.Y.: An Ultra Low-Power CMOS Transceiver Using Various Low-Power Techniques for LR-WPAN Applications. IEEE Transactions on Circuits and Systems–Part I: Regular Papers 59(2), 324–336 (2012)
65. Le, T., Mayaram, K., Fiez, T.: Efficient Far-Field Radio Frequency Energy Harvesting for Passively Powered Sensor Networks. IEEE Journal of Solid-State Circuits 43(5), 1287–1302 (2008)
66. Le, V.H., Han, S.K., Lee, J.S., Lee, S.G.: Current-Reused Ultra Low Power, Low Noise LNA+Mixer. IEEE Microwave and Wireless Components Letters 19(11), 755–757 (2009)
67. Le Roux, E., Scolari, N., Banerjee, B., Arm, C., Volet, P., Sigg, D., Heim, P., Perotto, J.F., Kaess, F., Raemy, N., Vouilloz, A., Ruffieux, D., Contaldo, M., Giroud, F., Severac, D., Morgan, M., Gyger, S., Monneron, C., Thanh-Chau, L., Henzelin, C., Peiris, V.: A 1V RF SoC with an 863-to-928MHz 400kb/s radio and a 32b Dual-MAC DSP core for Wireless Sensor and Body Networks. In: IEEE Int. Solid-State Circuits Conference Dig. Tech. Papers, pp. 464–465 (2010)
68. Lee, E.K.B., Kwon, H.M.: New baseband zero-crossing demodulator for wireless communications. I. Performance under static channel. In: IEEE Military Communications Conf., vol. 2, pp. 543–547 vol. 2 (1995)
69. Lee, H., Changkun, P., Songcheol, H.: A Quasi-Four-Pair Class-E CMOS RF Power Amplifier With an Integrated Passive Device Transformer. IEEE Transactions on Microwave Theory and Techniques 57(4), 752–759 (2009)
70. Lee, T.C., Chen, C.C.: A mixed-signal GFSK demodulator for Bluetooth. IEEE Transactions on Circuits and Systems–Part II: Express. Briefs 53(3), 197–201 (2006)
71. Lerdsitsomboon, W., Kenneth, K.O.: Technique for Integration of a Wireless Switch in a 2.4 GHz Single Chip Radio. IEEE Journal of Solid-State Circuits 46(2), 368–377 (2011)
72. Leung, L.L.K., Luong, H.C.: A 1-V 9.7-mW CMOS Frequency Synthesizer for IEEE 802.11a Transceivers. IEEE Transactions on Microwave Theory and Techniques 56(1), 39–48 (2008)
73. Li, P., Bashirullah, R.: A Wireless Power Interface for Rechargeable Battery Operated Medical Implants. IEEE Transactions on Circuits and Systems–Part II: Express. Briefs 54(10), 912–916 (2007)
74. Lin, S.H., Chiang, K.C., Chin, A., Yeh, F.S.: High-Density and Low-Leakage-Current MIM Capacitor Using Stacked TiO_2/ZrO_2 Insulators. Electron Device Letters, IEEE 30(7), 715–717 (2009)
75. Long, J.R.: Monolithic transformers for silicon RF IC design. IEEE Journal of Solid-State Circuits 35(9), 1368–1382 (2000)
76. Lu, J., Wang, N.Y., Chang, M.C.F.: A Compact and Low Power 5–10 GHz Quadrature Local Oscillator for Cognitive Radio Applications. IEEE Journal of Solid-State Circuits 47(5), 1131–1140 (2012)
77. Mandal, S., Sarpeshkar, R.: Low-Power CMOS Rectifier Design for RFID Applications. IEEE Transactions on Circuits and Systems–Part I: Regular Papers 54(6), 1177–1188 (2007)
78. Mangelsdorf, C.W.: A 400-MHz input flash converter with error correction. IEEE Journal of Solid-State Circuits 25(1), 184–191 (1990)

79. Manoli, Y.: Energy harvesting - from devices to systems. In: Proc. European Solid-State Circuits Conference, pp. 27–36 (2010)

80. Masmoudi, S., Ghazel, A., Loumeau, P.: Phase Data Converter Design for IEEE 802.15.4-based Wireless Receiver. In: IEEE Int. Conf. on Electronics, Circuits and Systems, pp. 955–958 (2007)

81. Mazzanti, A., Andreani, P.: A 1.4mW 4.90-to-5.65GHz Class-C CMOS VCO with an Average FoM of 194.5dBc/Hz. In: IEEE Int. Solid-State Circuits Conference Dig. Tech. Papers, pp. 474–629 (2008)

82. Mercier, P.P., Bhardwaj, M., Daly, D.C., Chandrakasan, A.P.: A 0.55V 16Mb/s 1.6mW non-coherent IR-UWB digital baseband with ±1ns synchronization accuracy. In: IEEE Int. Solid-State Circuits Conference Dig. Tech. Papers, pp. 252–253,253a (2009)

83. Mirzaei, A., Darabi, H., Leete, J.C., Xinyu, C., Juan, K., Yazdi, A.: Analysis and Optimization of Current-Driven Passive Mixers in Narrowband Direct-Conversion Receivers. IEEE Journal of Solid-State Circuits 44(10), 2678–2688 (2009)

84. Mirzaei, A., Heidari, M., Bagheri, R., Chehrazi, S., Abidi, A.: Injection-Locked Frequency Dividers based on Ring Oscillators with Optimum Injection for Wide Lock Range. In: Symp. on VLSI circuits, pp. 174–175 (2006)

85. Mirzaei, A., Heidari, M.E., Bagheri, R., Chehrazi, S., Abidi, A.A.: The Quadrature LC Oscillator: A Complete Portrait Based on Injection Locking. IEEE Journal of Solid-State Circuits 42(9), 1916–1932 (2007)

86. Mougiakakou, S.G., Bartsocas, C.S., Bozas, E., Chaniotakis, N., Iliopoulou, D., Kouris, I., Pavlopoulos, S., Prountzou, A., Skevofilakas, M., Tsoukalis, A., Varotsis, K., Vazeou, A., Zarkogianni, K., Nikita, K.S.: SMARTDIAB: A Communication and Information Technology Approach for the Intelligent Monitoring, Management and Follow-up of Type 1 Diabetes Patients. IEEE Transactions on Information Technology in Biomedicine 14(3), 622–633 (2010)

87. Oguey, H.J., Aebischer, D.: CMOS current reference without resistance. IEEE Journal of Solid-State Circuits 32(7), 1132–1135 (1997)

88. Ohnakado, T., Yamakawa, S., Murakami, T., Furukawa, A., Taniguchi, E., Ueda, H., Suematsu, N., Oomori, T.: 21.5-dBm power-handling 5-GHz transmit/receive CMOS switch realized by voltage division effect of stacked transistor configuration with depletion-layer-extended transistors (DETs). IEEE Journal of Solid-State Circuits 39(4), 577–584 (2004)

89. Okuni, H., Ito, R., Yoshida, H., Itakura, T.: A Direct Conversion Receiver with Fast-Settling DC Offset Canceller. In: IEEE 18th Int. Symp. on Personal, Indoor and Mobile Radio, Communications, pp. 1–5 (2007)

90. Oliveira, L.B., Fernandes, J.R., Filanovsky, I.M., Verhoeven, C.J., Silva, M.M.: Analysis and Design of Quadrature Oscillators. Springer, Analog circuits and signal processing series (2008)

91. Otis, B., Chee, Y.H., Rabaey, J.: A 400 μW-RX, 1.6mW-TX super-regenerative transceiver for wireless sensor networks. In: IEEE Int. Solid-State Circuits Conference Dig. Tech. Papers, pp. 396–606 Vol. 1 (2005)

92. Pamarti, S., Jansson, L., Galton, I.: A wideband 2.4-GHz delta-sigma fractional-N PLL with 1-Mb/s in-loop modulation. IEEE Journal of Solid-State Circuits 39(1), 49–62 (2004)

93. Papotto, G., Carrara, F., Finocchiaro, A., Palmisano, G.: A 90nm CMOS 5Mb/s crystal-less RF transceiver for RF-powered WSN nodes. In: IEEE Int. Solid-State Circuits Conference Dig. Tech. Papers, pp. 452–454 (2012)

94. Park, C., Kim, Y., Kim, H., Hong, S.: A 1.9-GHz Triple-Mode Class-E Power Amplifier for a Polar Transmitter. IEEE Microwave and Wireless Components Letters 17(2), 148–150 (2007)

95. Park, D., Cho, S.: A 1.8 V 900 μW 4.5 GHz VCO and Prescaler in 0.18 μm CMOS Using Charge-Recycling Technique. IEEE Microwave and Wireless Components Letters 19(2), 104–106 (2009)

96. Parvizi, M., Khodabakhsh, A., Nabavi, A.: Low-power high-tuning range CMOS ring oscillator VCOs. In: IEEE Int. Conf. on Semiconductor, Electronics, pp. 40–44 (2008)

97. Pelgrom, M.J.M., Duinmaijer, A.C.J., Welbers, A.P.G.: Matching properties of MOS transistors. IEEE Journal of Solid-State Circuits **24**(5), 1433–1439 (1989)

98. Perrott, M.H., Tewksbury, T.L.: I., Sodini, C.G.: A 27-mW CMOS fractional-N synthesizer using digital compensation for 2.5-Mb/s GFSK modulation. IEEE Journal of Solid-State Circuits **32**(12), 2048–2060 (1997)

99. Perumana, B.G., Mukhopadhyay, R., Chakraborty, S., Chang-Ho, L., Laskar, J.: A Low-Power Fully Monolithic Subthreshold CMOS Receiver With Integrated LO Generation for 2.4 GHz Wireless PAN Applications. IEEE Journal of Solid-State Circuits **43**(10), 2229–2238 (2008)

100. Pletcher, N.M., Gambini, S., Rabaey, J.: A 52 μW Wake-Up Receiver With -72 dBm Sensitivity Using an Uncertain-IF Architecture. IEEE Journal of Solid-State Circuits **44**(1), 269–280 (2009)

101. Popplewell, P., Karam, V., Shamim, A., Rogers, J., Roy, L., Plett, C.: A 5.2-GHz BFSK Transceiver Using Injection-Locking and an On-Chip Antenna. IEEE Journal of Solid-State Circuits **43**(4), 981–990 (2008)

102. Rai, S., Holleman, J., Pandey, J.N., Zhang, F., Otis, B.: A 500μW neural tag with $2\mu V_{rms}$ AFE and frequency-multiplying MICS/ISM FSK transmitter. In: IEEE Int. Solid-State Circuits Conference Dig. Tech. Papers, pp. 212–213,213a (2009)

103. Rai, S.S., Otis, B.P.: A 600 μW BAW-Tuned Quadrature VCO Using Source Degenerated Coupling. IEEE Journal of Solid-State Circuits **43**(1), 300–305 (2008)

104. Raja, M.K., Xuesong, C., Yan Dan, L., Zhao, B., Ben Choi, Y., Yuan, X.: A 18 mW Tx, 22 mW Rx transceiver for 2.45 GHz IEEE 802.15.4 WPAN in 0.18-μm CMOS. In: IEEE Asian Solid-State Circuits Conference Dig., pp. 41–44 (2010)

105. Razavi, B.: Design considerations for direct-conversion receivers. IEEE Transactions on Circuits and Systems–Part II: Analog and. Digital Signal Processing **44**(6), 428–435 (1997)

106. Razavi, B.: RF microelectronics. Prentice Hall PTR, Upper Saddle River, NJ (1998)

107. Razavi, B.: A study of injection locking and pulling in oscillators. IEEE Journal of Solid-State Circuits **39**(9), 1415–1424 (2004)

108. Redman-White, W., Leenaerts, D.M.W.: 1/f noise in passive CMOS mixers for low and zero IF integrated receivers. In: Proc. European Solid-State Circuits Conference, pp. 41–44 (2001)

109. Retz, G., Shanan, H., Mulvaney, K., O'Mahony, S., Chanca, M., Crowley, P., Billon, C., Khan, K., Quinlan, P.: A highly integrated low-power 2.4GHz transceiver using a direct-conversion diversity receiver in 0.18μm CMOS for IEEE802.15.4 WPAN. In: IEEE Int. Solid-State Circuits Conference Dig. Tech. Papers, pp. 414–415,415a (2009)

110. Reynaert, P.: A 2.45-GHz 0.13-μm CMOS PA With Parallel Amplification. IEEE Journal of Solid-State Circuits **42**(3), 551–562 (2007)

111. Reynaert, P., Steyaert, M.: RF Power Amplifiers for Mobile Communications. Springer, Dordrecht, The Netherlands (2006)

112. Rodriguez-Vazquez, A., Dominguez-Castro, R., Medeiro, F., Delgado-Restituto, M.: High resolution CMOS current comparators: design and applications to current-mode function generation. Analog Integrated Circuits and Signal Processing **7**(2), 149–165 (1995)

113. Rofougaran, A., Rael, J., Rofougaran, M., Abidi, A.: A 900 MHz CMOS LC-oscillator with quadrature outputs. In: IEEE Int. Solid-State Circuits Conference Dig. Tech. Papers, pp. 392–393 (1996)

114. Samadian, S., Hayashi, R., Abidi, A.A.: Demodulators for a zero-IF Bluetooth receiver. IEEE Journal of Solid-State Circuits **38**(8), 1393–1396 (2003)

115. Saputra, N., Long, J.R.: A Fully-Integrated, Short-Range, Low Data Rate FM-UWB Transmitter in 90 nm CMOS. IEEE Journal of Solid-State Circuits **46**(7), 1627–1635 (2011)

116. Schmidt, R., Norgall, T., Mörsdorf, J., Bernhard, J., von der Grün, T.: Body Area Network BAN - a Key Infrastructure Element for Patient-Centered Medical Applications. Biomedizinische Technik/Biomedical, Engineering 47(s1a), 365–368 (2002)

117. Seung-Wook, L., Kang-Yoon, L., Eunseok, S., Yeon-Jae, J., Hoisam, J., Jeong-Min, K., Hyeong-Jun, L., Jeong-Woo, L., Joonbae, P., Kyeongho, L., Soo-Ik, C., Deog-Kyoon, J., Wonchan, K.: A single-chip 2.4 GHz direct-conversion CMOS transceiver with GFSK modem for Bluetooth application. In: Symp. on VLSI circuits, pp. 245–246 (2001)

118. Shirvani, A., Su, D.K., Wooley, B.A.: A CMOS RF power amplifier with parallel amplification for efficient power control. IEEE Journal of Solid-State Circuits 37(6), 684–693 (2002)

119. Si, W.W., Weber, D., Abdollahi-Alibeik, S., MeeLan, L., Chang, R., Dogan, H., Haitao, G., Rajavi, Y., Luschas, S., Ozgur, S., Husted, P., Zargari, M.: A Single-Chip CMOS Bluetooth v2.1 Radio SoC. IEEE Journal of Solid-State Circuits 43(12), 2896–2904 (2008)

120. Solda, S., Caruso, M., Bevilacqua, A., Gerosa, A., Vogrig, D., Neviani, A.: A 5 Mb/s UWB-IR Transceiver Front-End for Wireless Sensor Networks in 0.13 μm CMOS. IEEE Journal of Solid-State Circuits 46(7), 1636–1647 (2011)

121. Srinivasan, R., Didem Zeliha, T., Sang Wook, P., Sanchez-Sinencio, E.: A Low-Power Frequency Synthesizer with Quadrature Signal Generation for 2.4 GHz Zigbee Transceiver Applications. In: Proc. IEEE Int. Symp. on Circuits and Systems, pp. 429–432 (2007)

122. Staszewski, R.B., Chih-Ming, H., Leipold, D., Balsara, P.T.: A first multigigahertz digitally controlled oscillator for wireless applications. IEEE Transactions on Microwave Theory and Techniques 51(11), 2154–2164 (2003)

123. Staszewski, R.B., Muhammad, K., Leipold, D., Chih-Ming, H., Yo-Chuol, H., Wallberg, J.L., Fernando, C., Maggio, K., Staszewski, R., Jung, T., Jinseok, K., John, S.: Irene Yuanying, D., Sarda, V., Moreira-Tamayo, O., Mayega, V., Katz, R., Friedman, O., Eliezer, O.E., de Obaldia, E., Balsara, P.T.: All-digital TX frequency synthesizer and discrete-time receiver for Bluetooth radio in 130-nm CMOS. IEEE Journal of Solid-State Circuits 39(12), 2278–2291 (2004)

124. Swaminathan, A., Wang, K.J., Galton, I.: A Wide-Bandwidth 2.4 GHz ISM Band Fractional-N PLL With Adaptive Phase Noise Cancellation. IEEE Journal of Solid-State Circuits 42(12), 2639–2650 (2007)

125. van der Tang, J., van de Ven, P., Kasperkovitz, D., van Roermund, A.: Analysis and design of an optimally coupled 5-GHz quadrature LC oscillator. IEEE Journal of Solid-State Circuits 37(5), 657–661 (2002)

126. Temporiti, E., Albasini, G., Bietti, I., Castello, R., Colombo, M.: A 700-kHz bandwidth $\Sigma\Delta$ fractional synthesizer with spurs compensation and linearization techniques for WCDMA applications. IEEE Journal of Solid-State Circuits 39(9), 1446–1454 (2004)

127. Thede, L.: Practical Analog and Digital Filter Design. Artech House, Inc. (2004)

128. Tiebout, M.: A 480 μW 2 GHz ultra low power dual-modulus prescaler in 0.25 μm standard CMOS. In: Proc. IEEE Int. Symp. on Circuits and Systems, vol. 5, pp. 741–744 vol. 5 (2000)

129. Tiebout, M.: Low Power VCO Design in CMOS. Springer Berlin Heidelberg, Berlin, Heidelberg (2006)

130. Tortori, P., Guermandi, D., Guermandi, M., Franchi, E., Gnudi, A.: Quadrature VCOs based on direct second harmonic locking: Theoretical analysis and experimental validation. International Journal of Circuit Theory and Applications 38(10), 1063–1086 (2010)

131. Tran, T.T.N., Boon, C.C., Do, M.A., Yeo, K.S.: Ultra-low power series input resonance differential common gate LNA. Electronics Letters 47(12), 703–704 (2011)

132. Vancorenland, P., Steyaert, M.S.J.: A 1.57-GHz fully integrated very low-phase-noise quadrature VCO. IEEE Journal of Solid-State Circuits 37(5), 653–656 (2002)

133. Vidojkovic, M., Huang, X., Harpe, P., Rampu, S., Zhou, C., Huang, L., van de Molengraft, J., Imamura, K., Büsze, B., Bouwens, F., Konijnenburg, M., Santana, J., Breeschoten, A., Huisken, J., Philips, K., Dolmans, G., de Groot, H.: A 2.4 GHz ULP OOK Single-Chip Transceiver for Healthcare Applications. IEEE Transactions on Biomedical Circuits and Systems 5(6), 523–534 (2011)

134. Visser, H.J., Reniers, A.C.F., Theeuwes, J.A.C.: Ambient RF Energy Scavenging: GSM and WLAN Power Density Measurements. In: European Microwave Conf., pp. 721–724 (2008)

135. Vullers, R.J., Visser, H.J., Op het Veld, B., Pop, V.: RF harvesting using antenna structures on foil. In: Int. Workshop on Micro and Nanotechnology for Power Generation and Energy Conversion Applications, pp. 209–212. Sendai, Japan (2008)

136. Vullers, R.J.M., Schaijk, R.V., Visser, H.J., Penders, J., Hoof, C.V.: Energy Harvesting for Autonomous Wireless Sensor Networks. IEEE Solid-State Circuits Mag. 2(2), 29–38 (2010)

137. Wang, J.H., Hsieh, H.H., Lu, L.H.: A 5.2-GHz CMOS T/R Switch for Ultra-Low-Voltage Operations. IEEE Transactions on Microwave Theory and Techniques **56**(8), 1774–1782 (2008)

138. Wang, X., Andreani, P.: Impact of mutual inductance and parasitic capacitance on the phase-error performance of CMOS quadrature VCOs. In: Proc. IEEE Int. Symp. on Circuits and Systems, vol. 1, pp. I-661-I-664 vol. 1 (2003)

139. Wang, X., Yu, Y., Busze, B., Pflug, H., Young, A., Huang, X., Zhou, C., Konijnenburg, M., Philips, K., De Groot, H.: A meter-range UWB transceiver chipset for around-the-head audio streaming. In: IEEE Int. Solid-State Circuits Conference Dig. Tech. Papers, pp. 450–452 (2012)

140. Wentzloff, D.D., Lee, F.S., Daly, D.C., Bhardwaj, M., Mercier, P.P., Chandrakasan, A.P.: Energy Efficient Pulsed-UWB CMOS Circuits and Systems. In: IEEE Int. Conf. on Ultra-Wideband, pp. 282–287 (2007)

141. Wong, A., Dawkins, M., Devita, G., Kasparidis, N., Katsiamis, A., King, O., Lauria, F., Schiff, J., Burdett, A.: A 1V 5mA multimode IEEE 802.15.6/bluetooth low-energy WBAN transceiver for biotelemetry applications. In: IEEE Int. Solid-State Circuits Conference Dig. Tech. Papers, pp. 300–302 (2012)

142. Wong, A.C.W., Kathiresan, G., Chan, C.K.T., Eljamaly, O., Omeni, O., McDonagh, D., Burdett, A.J., Toumazou, C.: A 1 V Wireless Transceiver for an Ultra-Low-Power SoC for Biotelemetry Applications. IEEE Journal of Solid-State Circuits **43**(7), 1511–1521 (2008)

143. Wu, W., Sanduleanu, M.A.T., Li, X., Long, J.R.: 17 GHz RF Front-Ends for Low-Power Wireless Sensor Networks. IEEE Journal of Solid-State Circuits **43**(9), 1909–1919 (2008)

144. Xiongchuan, H., Rampu, S., Xiaoyan, W., Dolmans, G., de Groot, H.: A 2.4GHz/915MHz 51μW wake-up receiver with offset and noise suppression. In: IEEE Int. Solid-State Circuits Conference Dig. Tech. Papers, pp. 222–223 (2010)

145. Yamagishi, A., Ugajin, M., Tsukahara, T.: A 1-V 2.4-GHz PLL synthesizer with a fully differential prescaler and a low-off-leakage charge pump. In: IEEE MTT-S Int. Microwave Symp. Dig., vol. 2, pp. 733–736 vol. 2 (2003)

146. Yeung Bun, C., Xiaojun, Y.: A 3.5-mW 2.45-GHz frequency synthesizer in 0.18μm CMOS. In: IEEE Int. Symp. Radio-Frequency Integration Technology, pp. 187–190 (2009)

147. Yousefi, R., Mason, R.: A 14-mW 2.4-GHz CMOS transceiver for short range wireless sensor applications. In: Proc. European Solid-State Circuits Conference, pp. 150–153 (2008)

148. Yuan, J., Svensson, C.: New single-clock CMOS latches and flipflops with improved speed and power savings. IEEE Journal of Solid-State Circuits **32**(1), 62–69 (1997)

149. Zahabi, A., Anis, M., Ortmanns, M.: 3.1GHz-3.8GHz integrated transmission line super-regeneration amplifier with degenerative quenching technique for impulse-FM-UWB transceiver. In: Proc. European Solid-State Circuits Conference, pp. 387–390 (2011)

150. Zhang, Y.P., Qiang, L., Wei, F.: Chew Hoe, A., He, L.: A Differential CMOS T/R Switch for Multistandard Applications. IEEE Transactions on Circuits and Systems–Part II: Express. Briefs **53**(8), 782–786 (2006)

151. Zheng, Y., Tong, Y., Ang, C.W., Xu, Y.P., Yeoh, W.G., Lin, F., Singh, R.: A CMOS Carrierless UWB Transceiver for WPAN Applications. In: IEEE Int. Solid-State Circuits Conference Dig. Tech. Papers, pp. 378–387 (2006)

Index

J. Masuch and M. Delgado-Restituto, *Ultra Low Power Transceiver for Wireless Body Area Networks*, Analog Circuits and Signal Processing, DOI: 10.1007/978-3-319-00098-5,
© Springer International Publishing Switzerland 2013

MIX
Papier aus verantwortungsvollen Quellen
Paper from responsible sources
FSC® C105338